Electron Diffraction and Structural Imaging II

Electron Diffraction and Structural Imaging II

Guest Editors

Partha Pratim Das
Arturo Ponce-Pedraza
Enrico Mugnaioli
Stavros Nicolopoulos

MDPI

Basel • Beijing • Wuhan • Barcelona • Belgrade • Novi Sad • Cluj • Manchester

Guest Editors

Partha Pratim Das
NanoMEGAS SPRL
Brussels
Belgium

Arturo Ponce-Pedraza
Department of Physics & Astronomy
University of Texas at San Antonio
San Antonio
USA

Enrico Mugnaioli
Department of Earth Sciences
University of Pisa
Pisa
Italy

Stavros Nicolopoulos
NanoMEGAS SPRL
Brussels
Belgium

Editorial Office
MDPI AG
Grosspeteranlage 5
4052 Basel, Switzerland

This is a reprint of the Special Issue, published open access by the journal *Symmetry* (ISSN 2073-8994), freely accessible at: https://www.mdpi.com/journal/symmetry/special_issues/Electron_Diffraction_Structural_Imaging_II.

For citation purposes, cite each article independently as indicated on the article page online and as indicated below:

Lastname, A.A.; Lastname, B.B. Article Title. *Journal Name* **Year**, *Volume Number*, Page Range.

ISBN 978-3-7258-5361-8 (Hbk)
ISBN 978-3-7258-5362-5 (PDF)
https://doi.org/10.3390/books978-3-7258-5362-5

Contents

About the Editors vii

Preface ix

Partha Pratim Das, Arturo Ponce-Pedraza, Enrico Mugnaioli and Stavros Nicolopoulos
Special Issue: Electron Diffraction and Structural Imaging—Volume II
Reprinted from: *Symmetry* **2025**, *17*, 1287, https://doi.org/10.3390/sym17081287 1

Iryna Andrusenko, Joseph Hitchen, Enrico Mugnaioli, Jason Potticary, Simon R. Hall and Mauro Gemmi
Two New Organic Co-Crystals Based on Acetamidophenol Molecules
Reprinted from: *Symmetry* **2022**, *14*, 431, https://doi.org/10.3390/sym14030431 5

Andrei Hernandez-Robles, David Romeu and Arturo Ponce
On the Mechanism Controlling the Relative Orientation of Graphene Bi-Layers
Reprinted from: *Symmetry* **2022**, *14*, 719, https://doi.org/10.3390/sym14040719 15

Christian Jandl, Gunther Steinfeld, Keyao Li, Pokka Ka Chuen Pang, Chun Lung Choi, Cengan Wang, et al.
Absolute Structure Determination of Chiral Zinc Tartrate MOFs by 3D Electron Diffraction
Reprinted from: *Symmetry* **2023**, *15*, 983, https://doi.org/10.3390/sym15050983 23

Yu-Jen Chou, Konstantin B. Borisenko, Partha Pratim Das, Stavros Nicolopoulos, Mauro Gemmi and Angus I. Kirkland
Influence of Precession Electron Diffraction Parameters and Energy Filtering on Reduced Density Function Analysis of Thin Amorphous Silica Films—Implications for Structural Studies
Reprinted from: *Symmetry* **2023**, *15*, 1291, https://doi.org/10.3390/sym15071291 36

Sara Passuti, Julien Varignon, Adrian David, Philippe Boullay
Scanning Precession Electron Tomography (SPET) for Structural Analysis of Thin Films along Their Thickness
Reprinted from: *Symmetry* **2023**, *15*, 1459, https://doi.org/10.3390/sym15071459 48

Khai-Nghi Truong, Sho Ito, Jakub M. Wojciechowski, Christian R. Göb, Christian J. Schürmann, Akihito Yamano, et al.
Making the Most of 3D Electron Diffraction: Best Practices to Handle a New Tool
Reprinted from: *Symmetry* **2023**, *15*, 1555, https://doi.org/10.3390/sym15081555 59

Przemysław Snopiński and Krzysztof Matus
Characterisation of Microstructure and Special Grain Boundaries in LPBF AlSi10Mg Alloy Subjected to the KoBo Extrusion Process
Reprinted from: *Symmetry* **2023**, *15*, 1634, https://doi.org/10.3390/sym15091634 76

Kevyn Gallegos-Moncayo, Justine Jean, Nicolas Folastre, Arash Jamali and Arnaud Demortière
Investigating Cathode Electrolyte Interphase Formation in NMC 811 Primary Particles through Advanced 4D-STEM ACOM Analysis
Reprinted from: *Symmetry* **2024**, *16*, 301, https://doi.org/10.3390/sym16030301 93

Taylan Örs, Irena Deroche, Corentin Chatelard, Mathias Dodin, Raquel Martinez-Franco, Alain Tuel and Jean-Louis Paillaud
Determination of Na^+ Cation Locations in Nanozeolite ECR-1 Using a 3D ED Method
Reprinted from: *Symmetry* **2024**, *16*, 477, https://doi.org/10.3390/sym16040477 107

About the Editors

Partha Pratim Das

Partha Pratim Das is an Application Scientist at NanoMEGAS SRL, Belgium. He earned his M.S. in Chemical Sciences from the Indian Institute of Science, Bangalore, and his Ph.D. in Materials Science from the University of Zurich in 2012, where he focused on structural disorder analysis using diffuse scattering from single-crystal X-ray data. Since then, at NanoMEGAS he has contributed to the development of precession-assisted 3D electron diffraction for solving complex nanostructures, electron pair distribution function mapping for amorphous materials, and novel methodologies for structural analysis. His research focuses on advanced electron diffraction approaches for beam-sensitive materials, in situ experiments, cultural heritage studies, and the investigation of amorphous and nanocrystalline systems. He has authored more than 50 publications with nearly 700 citations, co-organized over 30 international workshops including IUCr2017 in Hyderabad, and recently co-supervised a Ph.D. student from ITQ Valencia, Spain.

Arturo Ponce-Pedraza

Arturo Ponce-Pedraza is Professor and Department Chair of Physics and Astronomy at the University of Texas at San Antonio, where he leads the Structure Physics and Electron Microscopy group. He earned his MS in Solid State Physics from the University of Puebla, Mexico, and his PhD in Materials Science and Engineering from the University of Cadiz, Spain, in 2003. He carried out postdoctoral research at UNAM, Mexico, and at CNRS, France, between 2005 and 2007, focusing on electron crystallography of interfaces. From 2008 to 2011, he was Senior Scientist and head of the microscopy center at CIQA, Saltillo, Mexico. He joined UTSA in 2011 as Director of the Kleberg Advanced Microscopy Center and has been part of the faculty since 2013. He is the author of more than 120 peer-reviewed articles, including *Science*, *Nature*, and *ACS Nano*, and has supervised multiple doctoral and postdoctoral researchers. He is a member of MSA, MRS, and AMMM, and was inducted into the Mexican Academy of Science in 2018. His research spans electron diffraction, aberration-corrected TEM, holography, and in situ electron microscopy.

Enrico Mugnaioli

Enrico Mugnaioli is an Associate Professor of Mineralogy at the University of Pisa, Italy, where he manages the HR-TEM CISUP laboratory and provides TEM expertise for internal and external collaborations. He earned his Bachelor's and Master's degrees *cum laude* and his PhD in Earth Sciences from the University of Siena, where he applied electron crystallography methods for the characterization of serpentine minerals. From 2007 to 2014, he worked in Prof. U. Kolb's group at the University of Mainz, Germany, and was one of the first developers of three-dimensional electron diffraction (3DED). Between 2016 and 2022, he was Senior Scientific Collaborator at the Italian Institute of Technology@NEST in Pisa, developing low-dose diffraction protocols for beam-sensitive materials. He is currently focusing on the application of 3DED to different microcrystalline materials, among which are hybrid compounds, high-tech devices, nanoparticles, rare minerals, extraterrestrial samples, environmental specimens, porous materials and pharmaceuticals. He has authored more than 140 peer-reviewed papers with over 4400 citations.

Stavros Nicolopoulos

Stavros Nicolopoulos obtained his PhD in Materials Science in 1989 from the University of Grenoble, France, with a thesis on X-ray and neutron crystallography of magnetic structures. From 1991 to 1996, he was an Associate Professor at the Complutense University of Madrid, Spain, focusing on biomaterials research. Between 1996 and 2004, he worked at Philips Electron Optics as a TEM applications specialist and later served as President of FEI in Spain. Since 2004, he has been Director and Co-Founder of NanoMEGAS SRL, Brussels, a global leader in 4D-STEM precession electron diffraction instrumentation. His team developed the Digistar precession device and the ASTAR tool for phase and orientation imaging in TEM. He has authored more than 95 peer-reviewed publications with over 3300 citations and is co-inventor on five international patents. In 2004, he was appointed as a Consultant to the IUCr Commission on Electron Crystallography, and in 2011 he received the *Microscopy Today* Innovation Award for the ASTAR system, recognized as one of the ten most important inventions in electron microscopy that year. His research interests include advanced electron diffraction/crystallography, in situ TEM, and archaeometry.

Preface

Electron diffraction (ED) and structural imaging continue to advance as indispensable tools for structural science, offering atomic-scale insights across chemistry, physics, and materials science. This Reprint extends the themes of the first volume, presenting nine contributions that illustrate progress enabled by Cs correctors, direct detection cameras, beam precession, 4D-STEM, and 3DED. These methods now permit structural studies of nanocrystals, beam-sensitive compounds, and complex systems that are inaccessible to conventional X-ray techniques.

The scope includes organic charge-transfer co-crystals, graphene bilayers, chiral MOFs, amorphous silica films, perovskite thin films, additively manufactured alloys, lithium-ion cathodes, and zeolites. Featured advances span precession-assisted 3DED, graphene orientation mechanisms, absolute structural determination of chiral frameworks, electron reduced density function analysis of amorphous silica films, scanning precession tomography, and best practices for 3DED workflows. Applications address grain boundary engineering, cathode–electrolyte interphase studies, and the first complete structure determination of zeolite ECR-1.

Partha Pratim Das, Arturo Ponce-Pedraza, Enrico Mugnaioli, and Stavros Nicolopoulos
Guest Editors

symmetry MDPI

Editorial

Special Issue: Electron Diffraction and Structural Imaging—Volume II

Partha Pratim Das [1,*], Arturo Ponce-Pedraza [2], Enrico Mugnaioli [3] and Stavros Nicolopoulos [1]

1 NanoMEGAS SPRL, Rue Èmile Claus 49 bte 9, 1050 Brussels, Belgium; info@nanomegas.com
2 Department of Physics and Astronomy, The University of Texas at San Antonio, San Antonio, TX 78249, USA; arturo.ponce@utsa.edu
3 Department of Earth Sciences, University of Pisa, Via S. Maria, 53-56126 Pisa, Italy; enrico.mugnaioli@unipi.it
* Correspondence: partha@nanomegas.com

Following the success of the first edition of our Special Issue "Electron Diffraction and Structural Imaging", we present Volume II, featuring new and innovative contributions that further expand the scope and depth of this rapidly evolving field.

Driven by improved TEM hardware (e.g., Cs correctors, direct detector cameras) and novel techniques such as beam precession, 4D-STEM, and 3DED, electron diffraction (ED) and structural imaging enable atomic-scale analysis of a wide range of materials, including nanocrystals and beam-sensitive compounds. ED's low dosage and nanometric spatial resolution are ideal while X-ray methods fall short. Combining these methods with in situ sample holders, researchers can now study materials in real-time conditions, uncovering structural changes and symmetry evolution at the atomic level [1–10].

Volume II of this Special Issue features nine original contributions that showcase the versatility, depth, and continued innovation in the sphere of electron diffraction and structural imaging.

The first article, by Andrusenko et al., presents two new organic co-crystals formed via a simple solution growth method using paracetamol or metacetamol and 7,7,8,8-tetracyanoquinodimethane (TCNQ). These compounds belong to a class of organic charge-transfer complexes, with the acetamidophenol molecule acting as an electron donor and TCNQ as an acceptor. Due to the crystals' sub-micron sizes, precession-assisted 3D electron diffraction was employed for structural characterization. While the paracetamol–TCNQ structure was solved using standard direct methods, the metacetamol–TCNQ system required merging two datasets and applying simulated annealing due to its lower symmetry and data resolution. Both co-crystals have 1:1 stoichiometry, mixed-stack configuration, and non-centric P1 symmetry. Interestingly, they lack the strong hydrogen bonding seen in related orthocetamol-based co-crystals [11].

The second paper, by Hernández-Robles et al., investigates the relative orientations of rotated graphene bilayers (RGBs) grown via chemical vapor deposition. Using selected-area electron diffraction, the authors identified spontaneously preferred orientations that minimize lattice complexity based on possible Burgers vectors. They introduce the concept of secondary singular interfaces—non-singular orientations that still fulfill criteria of singularity based on their angular proximity to true singular states. These notable interfaces exhibit simpler displacement fields and reduced strain, suggesting that RGBs may reorient spontaneously toward such configurations. The study also provides a new explanation for the emergence of high-Σ interfaces, previously overlooked due to limitations in defining singularity [12].

The third paper, by Jandl et al. from Eldico Scientific, reports the absolute structural determination of anhydrous zinc (II) tartrate metal–organic frameworks (MOFs) with the chiral ligands [Zn (L-TAR)] and [Zn (D-TAR)] using electron diffraction on sub-micrometric crystals. The structures crystallize in the rare I222 space group, confirmed by dynamical refinement, which showed a clear distinction between enantiomorphs based on R-factor differences. These MOFs form dense 3D networks dominated by octahedral coordination, and similar structures were shown to be synthesizable using other divalent metals (Mg, Mn, Co, Ni, Cu). The chiral frameworks are described by Δ and Λ configurations, highlighting the potential of ED for the absolute structural determination of small crystals of chiral MOFs [13].

The fourth paper, by Chou et al., examines how precession angle, energy filtering, and sample thickness affect the structural analysis of amorphous SiO_2 thin films using electron reduced density functions. The authors find that while peak positions remain largely stable across conditions, peak intensities—and thus the derived coordination numbers—are significantly influenced by both precession and energy filtering. Notably, using small precession angles ($\leq 2^{\circ}$) and energy filtering yields coordination numbers for Si and O that better match the expected values of 4 and 2, particularly in thicker samples [14].

The fifth paper, by Passuti et al., introduces scanning precession electron tomography (SPET) as an advanced method for the accurate structural analysis of epitaxial perovskite thin films, particularly when the region of interest (ROI) is very small. By combining precession-assisted electron diffraction tomography (PEDT) with scanning over a defined area, the authors extracted spatially resolved intensities for precise structural refinement. A 35 nm $PrVO_3$ film on $SrTiO_3$ was used as a test case, revealing subtle structural variations across film thicknesses. It was demonstrated that SPET is a powerful alternative to traditional PEDT, capable of providing accurate structural data in ROIs as small as 10 nm [15].

The sixth paper, by Truong et al., shares practical insights from Rigaku's extensive experience with 3D electron diffraction (3D ED/MicroED) for solving structures of sub-micron single crystals. The authors highlight three key conclusions: (1) Cryo-transfer significantly improves data quality for hydrated compounds via preventing dehydration, as shown with trehalose dihydrate; (2) a streamlined workflow for dynamical diffraction enables reliable absolute structural determination for both small and large organic molecules (e.g., tyrosine and clarithromycin); (3) the crystal-to-detector distance must be optimized (even for small molecules such as cystine, longer distances can yield better results). These findings help define best practices across diverse sample types [16].

The seventh paper, by Snopiński et al., explores a novel grain boundary engineering (GBE) strategy for additively manufactured AlSi10Mg alloys using the KoBo extrusion method, which enables thermo-mechanical processing in a single step. Through EBSD and TEM analysis, the authors observed not only significant grain refinement but also an increased fraction of coincidence-site lattice (CSL) boundaries—especially low-Σ twin boundaries—indicating enhanced grain boundary character. The study links CSL formation to dynamic recrystallization, suggesting that KoBo extrusion offers a promising route to engineer tailored grain boundary networks in aluminum-based AM components. These results pave the way for next-generation high-performance alloys [17].

The eighth paper, by Gallegos-Moncayo et al., investigates cathode electrolyte interphase (CEI) formation in NMC 811 ($LiNi_{0.8}Mn_{0.1}Co_{0.1}O_2$), a key challenge in high-capacity lithium-ion batteries. Due to the proximity of the material's LUMO level to the HOMO of liquid electrolytes, oxidation reactions lead to surface degradation during charging. Using precession-assisted 4D-STEM-ACOM/ASTAR and STEM-EDX, the authors examined CEI composition at 4.3 V and 4.5 V, revealing a fluorine-rich layer that varied with voltage:

it consisted of LiF alone at 4.5 V and LiF + LiOH at 4.3 V. Despite LiF's reputation as a stable layer, it fails to prevent degradation in NMC 811. The study emphasizes the need for protective components, such as tailored additives or coatings, to enhance stability [18].

The ninth paper, by Örs et al., presents the first complete structural determination of zeolite ECR-1, an aluminosilicate with EON topology, previously inaccessible due to its submicron crystal size and stacking faults. Using a single nano-crystal with minimal defects, the authors performed precession electron diffraction (PED) at 103 K, achieving dynamical refinement with $Robs = 0.097$. The structure comprised 8.16 Na^+ ions across six crystallographic sites and ~four water molecules per unit cell. The material was hydrothermally synthesized with trioxane as a structure-directing agent, and a Monte Carlo simulation further validated the experimental cation and water distributions, marking significant progress in the structural study of faulted zeolites [19].

Conflicts of Interest: The authors declare no conflicts of interest.

References

1. Haider, M.; Uhlemann, S.; Schwan, E.; Rose, H.; Kabius, B.; Urban, K. Electron microscopy image enhanced. *Nature* **1998**, *392*, 768–769. [CrossRef]
2. Nogales, E. The development of cryo-EM into a mainstream structural biology technique. *Nat. Methods* **2016**, *13*, 24–27. [CrossRef] [PubMed]
3. Vincent, R.; Midgley, P.A. Double conical beam-rocking system for measurement of integrated electron diffraction intensities. *Ultramicroscopy* **1994**, *53*, 271–282. [CrossRef]
4. Ophus, C. Four-Dimensional Scanning Transmission Electron Microscopy (4D-STEM): From Scanning Nanodiffraction to Ptychography and Beyond. *Microsc. Microanal.* **2019**, *25*, 563–582. [CrossRef] [PubMed]
5. Gemmi, M.; Mugnaioli, E.; Gorelik, T.E.; Kolb, U.; Palatinus, L.; Boullay, P.; Hovmöller, S.; Abrahams, J.P. 3D Electron Diffraction: The Nanocrystallography Revolution. *ACS Cent. Sci.* **2019**, *5*, 1315–1329. [CrossRef] [PubMed]
6. Taheri, M.L.; Stach, E.A.; Arslan, I.; Crozier, P.A.; Kabius, B.C.; LaGrange, T.; Minor, A.M.; Takeda, S.; Tanase, M.; Wagner, J.B.; et al. Current status and future directions for in situ transmission electron microscopy. *Ultramicroscopy* **2016**, *170*, 86–95. [CrossRef] [PubMed]
7. Viladot, D.; Véron, M.; Gemmi, M.; Nicolopoulos, S.; Buffat, P.; Lefebvre, W.; Urban, K.; Midgley, P.A. Orientation and phase mapping in the transmission electron microscope using precession-assisted diffraction spot recognition: State-of-the-art results. *J. Microsc.* **2013**, *252*, 23–24. [CrossRef] [PubMed]
8. Cordero Oyonarte, E.; Rebecchi, L.; Gholam, S.; Faye Diouf, M.D.; Bigard, E.; Pralong, V.; Prestipino, C.; Kriegel, I.; Castellanos-Aliaga, A.; Hadermann, J.; et al. 3D Electron Diffraction on Nanoparticles: Minimal Size and Associated Dynamical Effects. *ACS Nano* **2025**, *19*, 20599–20612. [CrossRef] [PubMed]
9. Sun, Y.; Han, Y.; Zhou, D.; Galanis, A.S.; Gomez-Perez, A.; Wang, K.; Nicolopoulos, S.; Pérez-Garza, H.; Yang, Y. In Situ Nanometer-Resolution Strain and Orientation Mapping for Gas-Solid Reactions via Precession-Assisted Four-Dimensional Scanning Transmission Electron Microscopy. *Nano Today* **2025**, *64*, 102784. [CrossRef]
10. Demuth, T.; Kurzhals, P.; Ahmed, S.; Riewald, F.; Malaki, M.; Haust, J.; Beyer, A.; Janek, J.; Volz, K. Effect of a Two-Step Temperature-Swing Synthesis on Coarse-Grained $LiNiO_2$ Secondary Particles Characterized by Scanning Transmission Electron Microscopy. *Chem. Mater.* **2025**, *37*, 3993–4004. [CrossRef] [PubMed]
11. Andrusenko, I.; Hitchen, J.; Mugnaioli, E.; Potticary, J.; Hall, S.R.; Gemmi, M. Two New Organic Co-Crystals Based on Acetamidophenol Molecules. *Symmetry* **2022**, *14*, 431. [CrossRef]
12. Hernandez-Robles, A.; Romeu, D.; Ponce, A. On the Mechanism Controlling the Relative Orientation of Graphene Bi-Layers. *Symmetry* **2022**, *14*, 719. [CrossRef]
13. Jandl, C.; Steinfeld, G.; Li, K.; Pang, P.K.C.; Choi, C.L.; Wang, C.; Simoncic, P.; Williams, I.D. Absolute Structure Determination of Chiral Zinc Tartrate MOFs by 3D Electron Diffraction. *Symmetry* **2023**, *15*, 983. [CrossRef]
14. Chou, Y.-J.; Borisenko, K.B.; Das, P.P.; Nicolopoulos, S.; Gemmi, M.; Kirkland, A.I. Influence of Precession Electron Diffraction Parameters and Energy Filtering on Reduced Density Function Analysis of Thin Amorphous Silica Films—Implications for Structural Studies. *Symmetry* **2023**, *15*, 1291. [CrossRef]
15. Passuti, S.; Varignon, J.; David, A.; Boullay, P. Scanning Precession Electron Tomography (SPET) for Structural Analysis of Thin Films along Their Thickness. *Symmetry* **2023**, *15*, 1459. [CrossRef]

16. Truong, K.-N.; Ito, S.; Wojciechowski, J.M.; Göb, C.R.; Schürmann, C.J.; Yamano, A.; Del Campo, M.; Okunishi, E.; Aoyama, Y.; Mihira, T.; et al. Making the Most of 3D Electron Diffraction: Best Practices to Handle a New Tool. *Symmetry* **2023**, *15*, 1555. [CrossRef]
17. Snopiński, P.; Matus, K. Characterisation of Microstructure and Special Grain Boundaries in LPBF AlSi10Mg Alloy Subjected to the KoBo Extrusion Process. *Symmetry* **2023**, *15*, 1634. [CrossRef]
18. Gallegos-Moncayo, K.; Jean, J.; Folastre, N.; Jamali, A.; Demortière, A. Investigating Cathode Electrolyte Interphase Formation in NMC 811 Primary Particles through Advanced 4D-STEM ACOM Analysis. *Symmetry* **2024**, *16*, 301. [CrossRef]
19. Örs, T.; Deroche, I.; Chatelard, C.; Dodin, M.; Martinez-Franco, R.; Tuel, A.; Paillaud, J.-L. Determination of Na^+ Cation Locations in Nanozeolite ECR-1 Using a 3D ED Method. *Symmetry* **2024**, *16*, 477. [CrossRef]

symmetry

MDPI

Article

Two New Organic Co-Crystals Based on Acetamidophenol Molecules

Iryna Andrusenko [1], Joseph Hitchen [2], Enrico Mugnaioli [1], Jason Potticary [2], Simon R. Hall [2] and Mauro Gemmi [1,*]

1 Center for Materials Interfaces, Electron Crystallography, Istituto Italiano di Tecnologia, Viale Rinaldo Piaggio 34, 56025 Pontedera, Italy; iryna.andrusenko@iit.it (I.A.); enrico.mugnaioli@unipi.it (E.M.)

2 Complex Functional Materials Group, School of Chemistry, University of Bristol, Cantock's Close, Bristol BS8 1TS, UK; j.hitchen@bristol.ac.uk (J.H.); j.potticary@bristol.ac.uk (J.P.); simon.hall@bristol.ac.uk (S.R.H.)

* Correspondence: mauro.gemmi@iit.it

Abstract: Herein we present two new organic co-crystals obtained through a simple solution growth process based on an acetamidophenol molecule, either paracetamol or metacetamol, and on 7,7,8,8-tetracyanoquinodimethane (TCNQ). These co-crystals are part of a family of potential organic charge transfer complexes, where the acetamidophenol molecule behaves as an electron donor and TCNQ behaves as an electron acceptor. Due to the sub-micron size of the crystalline domains, 3D electron diffraction was employed for the structure characterization of both systems. Paracetamol-TCNQ structure was solved by standard direct methods, while the analysis of metacetamol-TCNQ was complicated by the low resolution of the available diffraction data and by the low symmetry of the system. The structure determination of metacetamol-TCNQ was eventually achieved after merging two data sets and combining direct methods with simulated annealing. Our study reveals that both paracetamol-TCNQ and metacetamol-TCNQ systems crystallize in a 1:1 stoichiometry, assembling in a mixed-stack configuration and adopting a non-centrosymmetric *P*1 symmetry. It appears that paracetamol and metacetamol do not form a strong structural scaffold based on hydrogen bonding, as previously observed for orthocetamol-TCNQ and orthocetamol-TCNB (1,2,4,5-tetracyanobenzene) co-crystals.

Keywords: electron diffraction; organic charge transfer complex; acetamidophenol co-crystals; structure determination; simulated annealing

1. Introduction

Symmetry is the true heart of crystallography. It is because of the symmetry that the description of a crystal passes from an Avogadro number of atoms to few atoms in the asymmetric unit. This simplifies enormously the problem of solving the crystal structure of an unknown phase. When the crystals become extremely small, in the range of few nanometers, the problem of symmetry determination becomes extremely challenging. The beautiful hints given by the crystal shape are hard to recognize in electron microscopy images of nanocrystals. The guide given by single-crystal diffraction data is not available when the size falls below a few microns. Then, the standard investigation technique becomes powder X-ray diffraction, which suffers from a reduction of dimensionality. In powder diffraction, the signal is one dimensional, while the symmetry is a three-dimensional quantity. Therefore, the determination of symmetry by powder diffraction can only be obtained indirectly through the profile fitting of the derived structural model.

In the last 15 years, since 2007, a new single-crystal diffraction technique became available for nanocrystals [1–3]. Because electrons interact stronger with matter than X-ray and can be focused in small bright beams, it is possible to record electron diffraction signals from crystals of a few hundreds of nanometers in any transmission electron microscope (TEM). This effect has been known since the invention of TEM, however, nobody thought to use the TEM as a single-crystal diffractometer until very recently. One reason was that the dynamical scattering, which spoils the simple linear relation between the diffraction intensities and the square modulus of the structure factors, was considered too strong for inorganic structures. On the other hand, organic structures are very beam sensitive so that only a few oriented patterns could be collected from them [4]. Still, electron diffraction had been successfully used for symmetry determination, especially in convergent beam electron diffraction (CBED) mode, but mainly on hard inorganic samples that are rather resistant to beam damage [5,6]. When scientists eventually tried to collect 3D electron diffraction (3D ED) data by recording a sequence of patterns while the crystal is rotated around the TEM goniometer axis [1], they discovered that the dynamical scattering is not so strong to completely hamper the structure solution based on kinematical approximation [7]. In particular, it was noticed that when reflections are integrated over their excitation error, the resulting 3D intensity data set can be considered quasi-kinematical.

The proper integration of reflection intensities was initially achieved by collecting each pattern of the sequence in precession mode, with the electron beam precessing on a cone surface having the vertex fixed on the sample plane [7]. The precession movement allows the Ewald sphere to sweep the reciprocal space around the actual orientation, integrating each nearby reflection over a certain excitation error. Later on, the same result was achieved by fine sampling the reciprocal space with small electrical beam tilts in the so-called rotation electron diffraction method [8]. Both of these acquisition protocols turned out to be very successful for the structure solution of unknown crystal structures [3].

The quasi-kinematical character of electron diffraction intensities guarantees the possibility to have valuable symmetry information, such as extinction conditions. However, the presence of residual dynamical scattering and experimental inaccuracies in the data collection always require them to be validated by a complete structure solution. This is particularly evident in the case of organic structures. Organic materials are beam sensitive; therefore, 3D ED methods should be performed under very low dose conditions. Nowadays this is achieved by the employment of a new generation of very sensitive detectors (direct electron detectors) and by the speeding of data collection with continuous and semi-automatic data collection while the crystal is rotating [9–13]. Yet, continuous rotation data collections on nanocrystalline materials is experimentally complicated. It is very difficult to have sufficient goniometer stability for keeping the electron beam on the same area for a wide angular range [14]. Moreover, the beam sensitivity of some materials is so pronounced to limit the number of patterns that can be collected before high-resolution reflections start to deteriorate. Thus, merging among several data sets may be required to have an adequate reciprocal space coverage [15]. Eventually, it is rather difficult to have a reliable a priori determination of the Laue class based on only intensity data before the correct phasing scheme is fully resolved.

With 3D ED data, we are generally compelled to perform several structure solution attempts with all the space groups compatible with the lattice geometry and with the detected extinction conditions. Interestingly, it is the structure solution that will reveal the real symmetry. A wrong symmetry will result in a structure model that is not chemically sound. For example, orthocetamol crystals exhibit a pseudo-tetragonal unit cell inside the precision of 3D ED data, but the symmetry had to be reduced to monoclinic in order to achieve the correct structure solution [16]. Similarly, the delta-polymorph of indomethacin was determined only after two independent molecules were introduced in the unit cell, obliterating the possibility of centrosymmetry [17].

Three-dimensional ED has been successfully used for the determination of nanocrystalline organic co-crystals [18,19]. In particular, Hitchen et al. [20] revealed the possibility of a new family of organic charge transfer (CT) complexes, based on rigid scaffold chains of orthocetamol molecules connected through hydrogen bonding and coupled with different planar acceptors. Here we report the structure solution of two other co-crystals based on acetamidophenol regioisomers (i.e., metacetamol and paracetamol) and 7,7,8,8-tetracyanoquinodimethane (TCNQ) (Figure 1). The structure determination of these two compounds was particularly challenging due to their beam sensitivity and triclinic symmetry, which reduced the overall data completeness. While the paracetamol-TCNQ co-crystal could be solved ab initio by direct methods, metacetamol-TCNQ co-crystal was determined by simulated annealing using a priori information about molecule connectivity. Both structures were subsequently optimized, verifying they were minima in the conformational energy landscape.

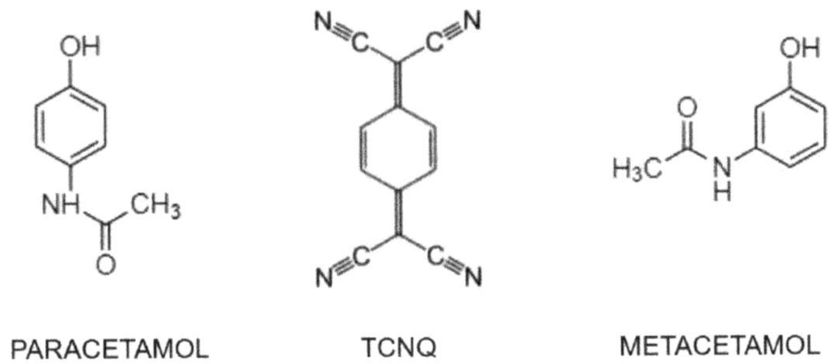

Figure 1. Molecular schemes of the three constituents used in this study.

2. Materials and Methods

2.1. *Crystallisation*

Paracetamol (>98% purity), metacetamol (>97%), and TCNQ (>98%) were obtained from Sigma-Aldrich and used as received without further purification. Donor and acceptor molecules were combined in a 1:1 M ratio in a minimum amount of anhydrous chloroform (Sigma-Aldrich, St. Louis, MO, USA, >99%) at room temperature. After 24 h, solutions were filtered at room temperature. After one week of slow evaporation, crystals of both paracetamol-TCNQ and metacetamol-TCNQ were removed from the solution and subsequently dried for analysis.

2.2. *TEM and 3D ED Structure Analysis*

High-angle annular dark-field scanning transmission electron microscopy (HAADF-STEM) imaging and 3D electron diffraction (3D ED) were performed with a Zeiss Libra 120 TEM operating at 120 kV and equipped with a thermionic LaB_6 source. A small amount of each co-crystal sample was gently crushed and directly loaded on a carbon-coated Cu TEM grid without any solvent or sonication. Three-dimensional ED was performed in STEM mode after defocusing the beam in order to have a parallel illumination on the sample, as described in Lanza et al. [21]. ED patterns were collected in Köhler parallel illumination with a beam size of about 150 nm in diameter, obtained with a 5 µm C2 condenser aperture and recorded by an ASI MEDIPIX single-electron detector [9]. This delivered virtually background-free diffraction patterns and allowed working with a very low electron dose.

For both samples, 3D ED data collections were performed at room temperature. No evidence of beam damage was ever observed, likely due to the extremely low dose rate. Three-dimensional ED data sets cover ranges 70–120°. Camera lengths of 180 mm were used, allowing for a resolution in real space of about 0.7 Å. Data were acquired in stepwise mode, with fixed steps of 1°. After each tilt, a diffraction pattern was acquired, and the crystal position was tracked by defocused STEM imaging. During stepwise experiments, the beam was precessed around the optical axis by an angle of 1° to improve reflection intensity integration [7]. Beam precession was performed using a Nanomegas Digistar P1000 device [22].

Three-dimensional ED data were analyzed with the software PETS 2.0 [23], using a low threshold during reflection search ($I/\sigma = 5$). Structure determinations were achieved by both standard direct methods (SDM) and simulated annealing (SA), as implemented in the software SIR2014 [24]. For SA structure determination, the models of single molecules were extracted from the Cambridge Structural Database [25]. Each co-former molecule was modeled as a unique fragment (Figure 1), where the atomic distances and coordination were known. No anti-bump restraint was used. Data were treated with a fully kinematical approximation, assuming that I_{hkl} was proportional to $|F_{hkl}|^2$. Least-squares structure refinements were performed based on the most complete acquisitions with the software SHELXL [26], using soft and rigid geometrical constraints. In the final refinement step, all hydrogen atoms were constrained in geometrically idealized positions. Atomic structures were visualized by the software VESTA [27].

2.3. Theoretical Calculations

Computational optimization was performed on both structures using CRYSTAL17 software [28] and calculated using the PBE0/6-31G level with 0 thermal component. Starting geometries were taken directly from experimentally determined models. Structures were compared to assess similarities between 3D ED-generated structures and the lowest energy structures obtained via calculation.

3. Results and Discussion

3.1. Structure Determination of Paracetamol-TCNQ Co-Crystal

Paracetamol-TCNQ crystallizes as yellow platelets up to 500 μm in size (Figure 2a). Such platelets are indeed agglomerates of much smaller crystalline domains, and this hinders the possibility of single-crystal X-ray diffraction experiments. Three-dimensional ED data were then recorded from six single-crystal fragments with a size less than 1 μm (Figure 2c). All six 3D ED data sets delivered a triclinic unit cell with approximate parameters $a = 7.20$ (14) Å, $b = 6.60$ (13) Å, $c = 9.20$ (18) Å, and $\alpha = 91.1$ (5)°, $\beta = 99.7$ (5)°, $\gamma = 86.1$ (5)°. A close look at 3D ED reconstructions revealed no hint of extinction features (Figure 3a–c). Considering the 1:1 ratio of paracetamol and TCNQ, such a cell would conveniently host only one pair of constituent molecules, which is consistent with the triclinic space group *P*1 (No. 1).

The structure of paracetamol-TCNQ co-crystal was determined ab initio by SDM using only the most complete 3D ED data set, i.e., the one acquired within the wider angular range and with better defined high-resolution reflections. All 27 non-hydrogen atoms of the structure were found by automatic routines after direct method phasing and both paracetamol and TCNQ molecules were immediately recognizable.

3.2. Structure Determination of Metacetamol-TCNQ Co-Crystal

Metacetamol-TCNQ appears as thin black platelets up to a few cm in length (Figure 2b). Again, each platelet is an agglomerate of much smaller crystalline domains. Three-dimensional ED data were recorded from five single-crystal fragments of different sizes (Figure 2d). All reconstructed 3D ED data delivered a triclinic primitive unit cell with parameters $a = 7.30$ (15) Å, $b = 9.40$ (19) Å, $c = 9.80$ (20) Å and $\alpha = 106.0$ (5)°, $\beta = 93.4$ (5)°, $\gamma = 92.3$ (5)°. No extinction features were detected (Figure 3d–f), and cell volume appeared consistent with only one pair of constituent molecules.

Diffraction data from metacetamol-TCNQ turned out to be of lower quality than the ones collected from paracetamol-TCNQ. Indeed, no structure solution could be achieved using a single 3D ED data set. Two data sets were therefore merged with a scale factor based on the strongest reflections (arguably the ones whose intensity values were relatively less affected by dynamical effects, experimental inaccuracies, and background).

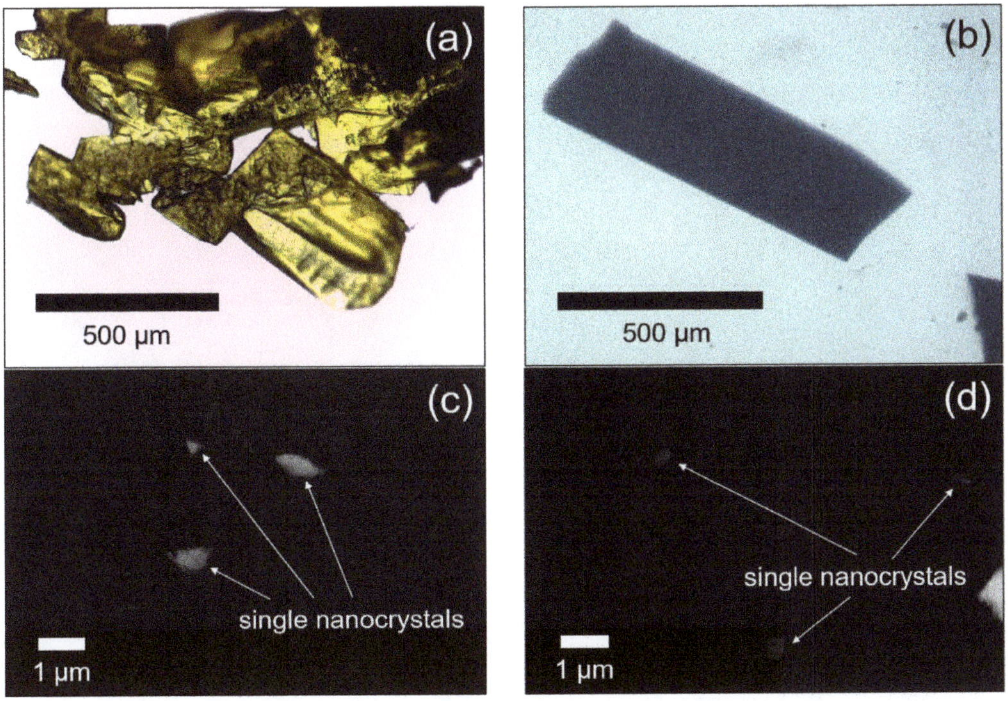

Figure 2. Light microscopy images of paracetamol-TCNQ (**a**) and metacetamol-TCNQ (**b**) platelets. HAADF-STEM images of paracetamol-TCNQ nanocrystals (**c**) and metacetamol-TCNQ nanocrystals (**d**) selected for 3D ED data collection.

Using this merged data set, the structure of metacetamol-TCNQ could be solved by SDM in space group *P*1 (No. 1). The maxima of the resulting potential map were poorly defined. One atom of metacetamol ring was not spotted by automatic routines, while the TCNQ molecule was largely incomplete, especially around the cyanide group, where atoms were relatively close (C≡N bond distance is expected in the range 1.16–1.11 Å). To some extent, the resolution of the cyanide group was also complicate in the previously reported orthocetamol-TCNQ co-crystal, where considerably better data were available [20].

The structure of metacetamol-TCNQ was independently confirmed by SA method. This global optimization method has been already applied to poor quality 3D ED data for the determination of organic structures [8,17,19,29,30]. SA method appears particularly suitable for this case, because TCNQ molecule is rigid and acetamidophenols have only one free torsion angle.

3.3. Structure Refinement and Energy Minimisation

Structure models of paracetamol-TCNQ and metacetamol-TCNQ were eventually least-squares refined against 3D ED data using SHELXL [26]. Geometrical ties were added stepwise to check the stability of the models. All hydrogen atoms were generated in idealized positions during the last refinement cycle. More details about the structures determination and refinement are listed in Table 1. Final models for paracetamol-TCNQ and metacetamol-TCNQ crystal structures are shown in Figure 4. Related CIF files have been uploaded in the CCDC database, with deposition numbers 2147971 and 2148006, respectively.

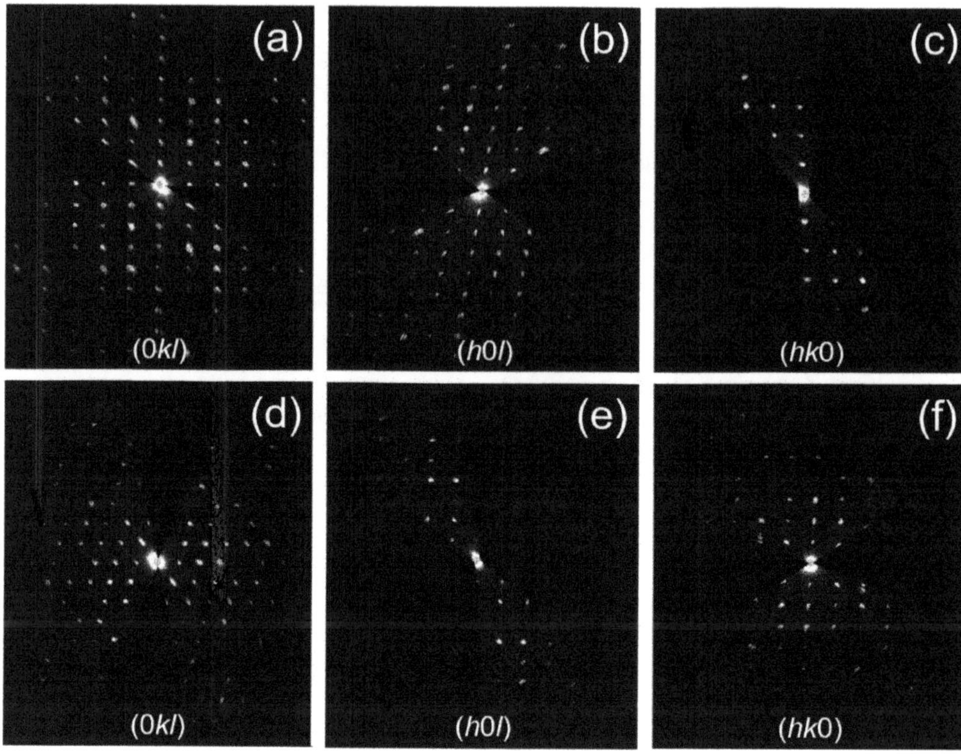

Figure 3. Selected planar cuts of the 3D ED reconstruction from paracetamol-TCNQ (**a**–**c**) and metacetamol-TCNQ (**d**–**f**) co-crystals.

Table 1. Selected parameters from structures determination and refinement.

	Paracetamol-TCNQ	**Metacetamol-TCNQ**
ab initio structure determination by SIR2014		
Data resolution (Å)	0.8	0.9
Sampled reflections (No.)	2296	2763
Independent reflections (No.)	1251	1403
Independent reflection coverage (%)	71	76
Global thermal factor U_{iso} (Å^2)	0.02820	0.03214
R_{int} (%)	20.19	16.01
R_{SIR} (%)	27.60	29.82
structure refinement by SHELXL		
Data resolution (Å)	0.9	0.9
Reflections total (No.)	1492	2763
Reflections > 4σ (No.)	648	1449
$R1_{4\sigma}$ (%)	37.01	34.60
$R1_{all}$ (%)	39.46	46.28
GooF	2.161	3.850

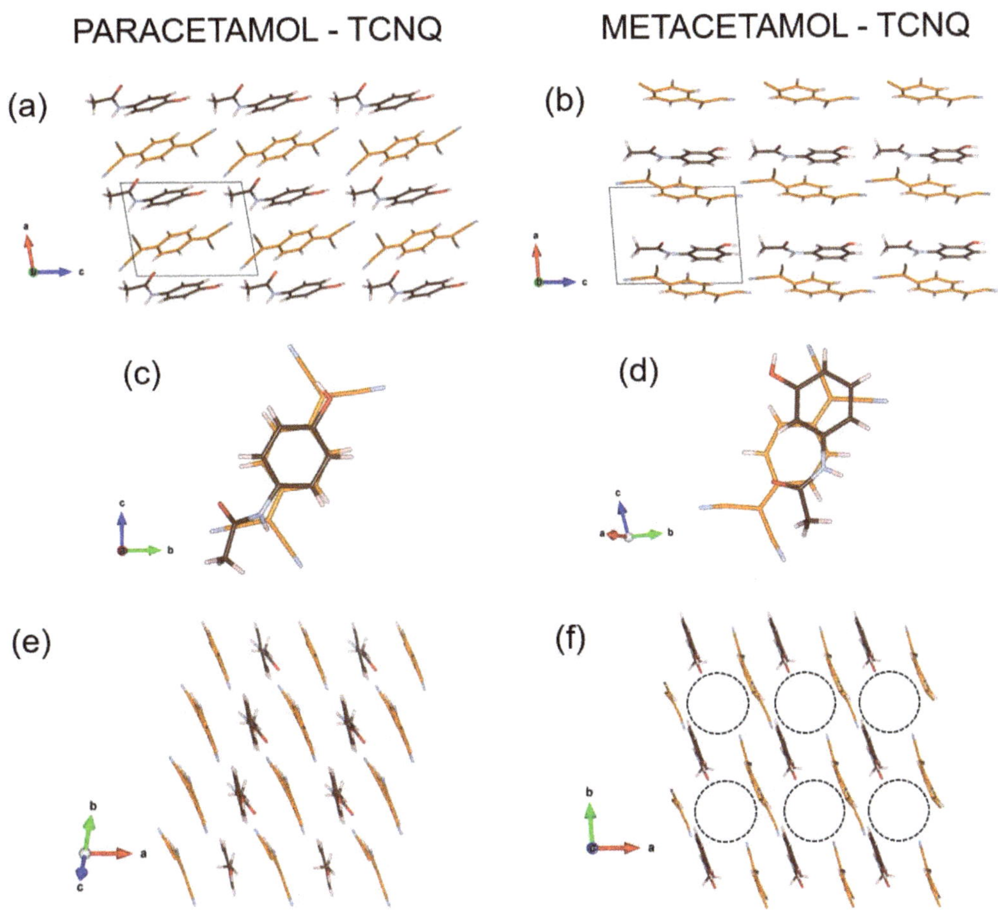

Figure 4. Paracetamol-TCNQ (**a**,**c**,**e**) and metacetamol-TCNQ (**b**,**d**,**f**) structures viewed along selected projections. Carbon atoms of paracetamol and metacetamol molecules are in brown, carbon atoms of TCNQ molecule are in orange, nitrogen atoms are always in light blue, oxygen atoms are in red, and hydrogen atoms are in light grey. The [001] channel in metacetamol-TCNQ is emphasized by a dashed circle (diameter about 6 Å).

Computational optimization was performed on both structures using CRYSTAL17 software [28], as previously done for olanzapine-phenol co-crystal [19]. A comparison of the experimental and calculated structures shows only minor differences in the hydrogen-bond lengths or relative ring angles, while the overall structure remains virtually intact. This confirms that the models determined experimentally based on 3D ED data are indeed structural energy minima (Figure 5).

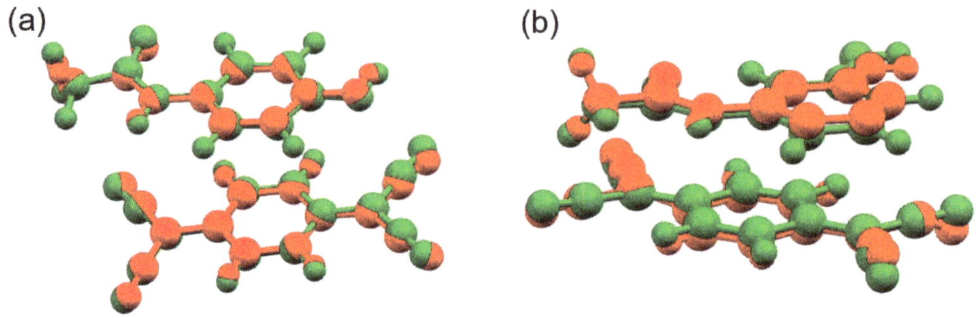

Figure 5. Overlay of the experimental (red) and calculated (green) structures of paracetamol-TCNQ (**a**) and metacetamol-TCNQ (**b**) co-crystals.

3.4. Structure Description

Paracetamol-TCNQ and metacetamol-TCNQ structures show a mixed-stack motif, as observed in many organic CT co-crystals [20,31,32]. In paracetamol-TCNQ, molecules of paracetamol and TCNQ are piled along their 6-carbon rings, following an 'eclipsed' arrangement such as in orthocetamol-TCNQ [20] (Figure 4c). In metacetamol-TCNQ, molecules show a looser piling, and the TCNQ ring is centered on the amide-phenol junction of the metacetamol molecule (Figure 4d). This packing generates a large channel along [001], with a diameter of more than 6 Å (Figure 4f).

Both paracetamol-TCNQ and metacetamol-TCNQ show no evidence of hydrogen bonding between acetamidophenol molecules. This is the most striking difference with orthocetamol-TCNQ co-crystal, whose structure is based on rigid scaffold chains of orthocetamol molecules [20]. Eventually, metacetamol-TCNQ and paracetamol-TCNQ appear more similar to conventional organic CT co-crystals, based on flat and loosely connected molecules.

We also point out the drastic reduction of symmetry from orthocetamol-TCNQ to paracetamol-TCNQ and metacetamol-TCNQ co-crystals. The former crystallizes in monoclinic space group *Pc*, with four couples of constituent molecules hosted in the asymmetric unit. Instead, both compounds reported in this paper crystallize in the triclinic space group *P*1, characterized by the only translational symmetry. Because the three compounds were obtained following a comparable crystallization route, such reduction of symmetry is dictated by the different acetamidophenol molecules. In particular, structure differences appear mostly related to the tendency of orthocetamol to form stiff backbone chains [16]. Interestingly, all acetamidophenol-TCNQ co-cocrystals reported up to date are non-centrosymmetric.

4. Conclusions

The structures of paracetamol-TCNQ and metacetamol-TCNQ co-crystals were solved using 3D electron diffraction despite remarkable experimental difficulties, such as small crystal size, beam sensitivity, and low symmetry. This result would be impossible to obtain with conventional single-crystal X-ray methods and hardly feasible even with cutting-edge powder X-ray facilities due to severe peak broadening and overlapping.

This work completes the study of acetamidophenol-TCNQ co-cocrystals produced through the simple evaporation from an anhydrous chloroform solution. While orthocetamol-TCNQ showed a more symmetric arrangement based on rigid scaffold chains of orthocetamol molecules [20], paracetamol-TCNQ and metacetamol-TCNQ crystallize in the lowest symmetric space group *P*1 and exhibit motifs more similar to typical mixed-stack organic CT co-crystals. This example emphasizes how much structures and symmetry of co-crystals may be affected by seemingly minor differences in the molecular configuration of their constituents, such as the ones that occur among the three acetamidophenol regioisomers.

Author Contributions: Conceptualization, S.R.H. and M.G.; investigation, I.A., J.H., E.M. and J.P.; data curation, E.M.; writing—original draft preparation, I.A., J.H., E.M., J.P. and M.G.; writing—review and editing, I.A., S.R.H. and M.G.; visualization, I.A. and J.H.; supervision, S.R.H. and M.G.; funding acquisition, S.R.H. and M.G. All authors have read and agreed to the published version of the manuscript.

Funding: This research was funded by FELIX project (Por CREO FESR 2014-2020 action), Engineering and Physical Sciences Research Council UK (grants EP/G036780/1 and EP/L015544/1), European Union's Horizon 2020 Research and Innovation program (grant No. 736899).

Institutional Review Board Statement: Not applicable.

Informed Consent Statement: Not applicable.

Data Availability Statement: CCDC 2147971 contains the supplementary crystallographic data for paracetamol-TCNQ co-crystal and CCDC 2148006 contains the supplementary crystallographic data for metacetamol-TCNQ co-crystal. These data can be obtained free of charge via www.ccdc.cam.ac.uk/data_request/cif (accessed on 14 February 2022), or by emailing data_request@ccdc.cam.ac.uk, or by contacting The Cambridge Crystallographic Data Centre, 12 Union Road, Cambridge CB2 1EZ, UK; fax: +44 1223 336 033.

Acknowledgments: I.A., E.M. and M.G. acknowledge the Regione Toscana for funding the purchase of the Timepix. J.H., J.P. and S.R.H. acknowledge MagnaPharm for a collaborative research, the Bristol Centre for Functional Nanomaterials and the Centre for Doctoral Training in Condensed Matter Physics.

Conflicts of Interest: The authors declare no conflict of interest.

References

1. Kolb, U.; Gorelik, T.; Kübel, C.; Otten, M.T.; Hubert, D. Towards automated diffraction tomography: Part I—Data acquisition. *Ultramicroscopy* **2007**, *107*, 507–513. [CrossRef] [PubMed]
2. Shi, D.; Nannenga, B.L.; Iadanza, M.G.; Gonen, T. Threedimensional electron crystallography of protein microcrystals. *eLife* **2013**, *2*, e01345. [CrossRef]
3. Gemmi, M.; Mugnaioli, E.; Gorelik, T.E.; Kolb, U.; Palatinus, L.; Boullay, P.; Hovmöller, S.; Abrahams, J.P. 3D electron diffraction: The nanocrystallography revolution. *ACS Cent. Sci.* **2019**, *5*, 1315–1329. [CrossRef] [PubMed]
4. Dorset, D.L. *Structural Electron Crystallography*; Plenum Press: New York, NY, USA, 1995.
5. Steeds, J.W. Convergent beam electron diffraction. In *Introduction to Analytical Electron Microscopy*; Hren, J.J., Goldstein, J.I., Joy, D.C., Eds.; Springer: Boston, MA, USA, 1979.
6. Zuo, J.M.; Spence, J.C.H. Automated structure factor refinement from convergent-beam patterns. *Ultramicroscopy* **1991**, *35*, 185–186. [CrossRef]
7. Mugnaioli, E.; Gorelik, T.; Kolb, U. "Ab Initio" structure solution from electron diffraction data obtained by a combination of automated diffraction tomography and precession technique. *Ultramicroscopy* **2009**, *109*, 758–765. [CrossRef] [PubMed]
8. Zhang, D.; Oleynikov, P.; Hovmöller, S.; Zou, X. Collecting 3D electron diffraction data by the rotation method. *Z. Kristallogr.* **2010**, *225*, 94–102. [CrossRef]
9. Nederlof, I.; van Genderen, E.; Li, Y.-W.; Abrahams, J.P. A Medipix quantum area detector allows rotation electron diffraction data collection from submicrometre three-dimensional protein crystals. *Acta Crystallogr. D* **2013**, *69*, 1223–1230. [CrossRef]
10. Nannenga, B.L.; Shi, D.; Leslie, A.G.W.; Gonen, T. High-resolution structure determination by continuous-rotation data collection in MicroED. *Nat. Methods* **2014**, *11*, 927–930. [CrossRef]
11. Gruene, T.; Wennmacher, J.T.C.; Zaubitzer, C.; Holstein, J.J.; Heidler, J.; Fecteau-Lefebvre, A.; De Carlo, S.; Müller, E.; Goldie, K.N.; Regeni, I.; et al. Rapid structure determination of microcrystalline molecular compounds using electron diffraction. *Angew. Chem. Int. Ed.* **2018**, *57*, 16313–16317. [CrossRef]
12. Jones, C.G.; Martynowycz, M.W.; Hattne, J.; Fulton, T.J.; Stoltz, B.M.; Rodriguez, J.A.; Nelson, H.M.; Gonen, T. The cryoEM method microED as a powerful tool for small molecule structure determination. *ACS Cent. Sci.* **2018**, *4*, 1587–1592. [CrossRef]
13. Huang, Z.; Grape, S.E.; Li, J.; Inge, A.K.; Zou, X. 3D electron diffraction as an important technique for structure elucidation of metal-organic frameworks and covalent organic frameworks. *Coordin. Chem. Rev.* **2021**, *427*, 213583. [CrossRef]
14. Gemmi, M.; Lanza, A.E. 3D electron diffraction techniques. *Acta Crystallogr. B* **2019**, *75*, 495–504. [CrossRef] [PubMed]
15. Ge, M.; Yang, T.; Wang, Y.; Carraro, F.; Liang, W.; Doonan, C.; Falcaro, P.; Zheng, H.; Zou, X.; Huang, Z. On the completeness of three-dimensional electron diffraction data for structural analysis of metal–organic frameworks. *Faraday Discuss.* **2021**, *231*, 66–80. [CrossRef] [PubMed]
16. Andrusenko, I.; Hamilton, V.; Mugnaioli, E.; Lanza, A.; Hall, C.; Potticary, J.; Hall, S.R.; Gemmi, M. The crystal structure of orthocetamol solved by 3D electron diffraction. *Angew. Chem. Int. Ed.* **2019**, *58*, 10919–10922. [CrossRef]

17. Andrusenko, I.; Hamilton, V.; Lanza, A.E.; Hall, C.L.; Mugnaioli, E.; Potticary, J.; Buanz, A.; Gaisford, S.; Piras, A.M.; Zambito, Y.; et al. Structure determination, thermal stability and dissolution rate of δ-indomethacin. *Int. J. Pharm.* **2021**, *608*, 121067. [CrossRef]
18. Brázda, P.; Palatinus, L.; Babor, M. Electron diffraction determines molecular absolute configuration in a pharmaceutical nanocrystal. *Science* **2019**, *364*, 667–669. [CrossRef]
19. Andrusenko, I.; Potticary, J.; Hall, S.R.; Gemmi, M. A new olanzapine cocrystal obtained from volatile deep eutectic solvents and determined by 3D electron diffraction. *Acta Crystallogr. B* **2020**, *76*, 1036–1044. [CrossRef]
20. Hitchen, J.; Andrusenko, I.; Hall, C.L.; Mugnaioli, E.; Potticary, J.; Gemmi, M.; Hall, S.R. Organic cocrystals of TCNQ and TCNB based on an orthocetamol backbone solved by three-dimensional electron diffraction. *Cryst. Growth Des.* **2022**, *22*, 1155–1163. [CrossRef]
21. Lanza, A.; Margheritis, E.; Mugnaioli, E.; Cappello, V.; Garau, G.; Gemmi, M. Nanobeam precession-assisted 3D electron diffraction reveals a new polymorph of hen egg-white lysozyme. *IUCrJ* **2019**, *6*, 178–188. [CrossRef]
22. Vincent, R.; Midgley, P.A. Double conical beam-rocking system for measurement of integrated electron diffraction intensities. *Ultramicroscopy* **1994**, *53*, 271–282. [CrossRef]
23. Palatinus, L.; Brázda, P.; Jelínek, M.; Hrdá, J.; Steciuk, G.; Klementová, M. Specifics of the data processing of precession electron diffraction tomography data and their implementation in the program PETS2.0. *Acta Crystallogr. B* **2019**, *75*, 512–522. [CrossRef] [PubMed]
24. Burla, M.C.; Caliandro, R.; Carrozzini, B.; Cascarano, G.L.; Cuocci, C.; Giacovazzo, C.; Mallamo, M.; Mazzone, A.; Polidori, G. Crystal structure determination and refinement via SIR2014. *J. Appl. Crystallogr.* **2015**, *48*, 306–309. [CrossRef]
25. Groom, C.R.; Bruno, I.J.; Lightfoot, M.P.; Ward, S.C. The Cambridge structural database. *Acta Crystallogr. B* **2016**, *72*, 171–179. [CrossRef] [PubMed]
26. Sheldrick, G.M. Crystal structure refinement with SHELXL. *Acta Crystallogr. C* **2015**, *71*, 3–8. [CrossRef] [PubMed]
27. Momma, K.; Izumi, F.J. VESTA 3 for three-dimensional visualization of crystal, volumetric and morphology data. *J. Appl. Crystallogr.* **2011**, *44*, 1272–1276. [CrossRef]
28. Dovesi, R.; Erba, A.; Orlando, R.; Zicovich-Wilson, C.M.; Civalleri, B.; Maschio, L.; Rérat, M.; Casassa, S.; Baima, J.; Salustro, S.; et al. Quantum-mechanical condensed matter simulations with CRYSTAL. *WIREs Comput. Mol. Sci.* **2018**, *8*, e1360. [CrossRef]
29. Das, P.P.; Mugnaioli, E.; Nicolopoulos, S.; Tossi, C.; Gemmi, M.; Galanis, A.; Borodi, G.; Pop, M.M. Crystal structures of two important pharmaceuticals solved by 3D precession electron diffraction tomography. *Org. Process Res. Dev.* **2018**, *22*, 1365–1372. [CrossRef]
30. Feyand, M.; Mugnaioli, E.; Vermoortele, F.; Bueken, B.; Dieterich, J.M.; Reimer, T.; Kolb, U.; de Vos, D.; Stock, N. Automated diffraction tomography for the structure elucidation of twinned, sub-micrometer crystals of a highly porous, catalytically active bismuth metal–organic framework. *Angew. Chem. Int. Ed.* **2012**, *51*, 10373–10376. [CrossRef]
31. Yu, W.; Wang, X.-Y.; Li, J.; Li, Z.-T.; Yan, Y.-K.; Wang, W.; Pei, J. A photoconductive charge-transfer crystal with mixed-stacking donor–acceptor heterojunctions within the lattice. *Chem. Commun.* **2013**, *49*, 54–56. [CrossRef]
32. Zhu, L.; Yi, Y.; Fonari, A.; Corbin, N.S.; Coropceanu, V.; Brédas, J.-L. Electronic properties of mixed-stack organic charge-transfer crystals. *J. Phys. Chem. C* **2014**, *118*, 14150–14156. [CrossRef]

symmetry

MDPI

Article

On the Mechanism Controlling the Relative Orientation of Graphene Bi-Layers

Andrei Hernandez-Robles [1], David Romeu [2] and Arturo Ponce [1,3,*]

1 Department of Physics and Astronomy, University of Texas at San Antonio, One UTSA Circle, San Antonio, TX 78249, USA; andrei.hernadezrobles@utsa.edu
2 Instituto de Física, Universidad Nacional Autónoma de México, Apartado Postal 20-364, Mexico City 01000, Mexico; romeu@fisica.unam.mx
3 Instituto Nacional de Astrofísica, Óptica y Electrónica, Puebla 72840, Mexico
* Correspondence: arturo.ponce@utsa.edu

Abstract: We have measured the relative orientation of rotated graphene bi-layers (RGBs) deposited by chemical vapor deposition and found that there are spontaneously occurring preferred orientations. Measurements were performed using selected area electron diffraction patterns on various regions of the RGBs. These orientations minimize the complexity of the lattice defined by the set of all possible Burgers vectors. By using a precise definition of singularity, we have been able to show that all non-singular preferred orientations are special in the sense that their angular distance $\Delta\theta$ to the closest singular orientation also complies with the definition of singularity. Our results show that these special interfaces, named secondary singular interfaces,have simpler displacement fields compared to other non-singular RGBs, implying that interfacial dislocations have fewer Burgers vectors to choose from. Since all observed orientations were found to be either singular or secondary singular, we found evidence that RGBs starting out with rotation angles far from singular orientations re-orient themselves into a nearby secondary singular state in order to simplify their strain fields. Secondary singular orientations also account for the spontaneous formation of high Σ interfaces, although the lack of a precise definition of singularity caused them to remain unnoticed.

Keywords: crystallography of interfaces; electron diffraction; coincidence site lattice; dichromatic patterns

1. Introduction

Two-dimensional crystals, such as graphene and metal dichalcogenides (MX_2), have risen in the interest of the scientific community due to their novel technological applications. Physical properties of confined 2D crystals can also be controlled as a function of the number of their assembled monolayers, e.g., a bilayer graphene is a tunable bandgap and its charge carrier mobility can be controlled as well [1]. The MoS_2 monolayer is a direct band-gap semiconductor, unlike its bulk nature at 1.2 eV [2,3], and, as a consequence, a remarkable increase in luminescence is experimentally measured [4]. In contrast with the extended studies of graphene 2D sheets, metal dichalcogenides exhibit a band gap shift from indirect to direct as a function of the number of S-Mo-S slabs (N), and increasing the band gap from the bulk value up to 0.6 eV for a single MoS_2 layer [2,5].Those crystals with few layers can be stacked up directly, one on top of another, with no rotation; however, properties such as those mentioned above are affected, depending on the rotation angles [6].In this way, interfacial phenomena and the analysis of the preferred orientations of the 2D-bilayer crystals are fundamental topics for their technological applications. Given that the electronic properties of rotated graphene bi-layers, hereafter referred to as rotational graphene bi-layers (RGBs), depend on the rotation angle, in this work we inquire about the possibility that there exist special orientations with low interfacial energies indicative of possible special structural and electronic properties. These rotation angles between bi-layers are, nowadays, referred as magic angles and are responsible to unusual insulating and superconductivity properties for lower twisting angles in graphene [7,8]. Recently, a complementary continuum model has been reported, where the Fermi velocities of the moiré Dirac points not only vanish, but the bands become entirely flat at a set of punctual magic angles found at a periodicity of $\Delta\alpha \simeq 3/2$ [9].

In crystalline materials, short period interfaces identified by a small index number Σ (the ratio of the volume of the unit cell of the coincidence site lattice CSL [10] to that of the unit cell of the parent crystals) are known to be preferred, since the now classical experiments of Chaudhari and Matthews [11], who observed that small Σ interfaces occur preferentially in bi-crystals spontaneously formed during the condensation of MgO and CdO smoke. They also observed, however, that a number of bi-crystals with rotation angles lead to large values of Σ. The reason why these angles have preferred orientations has no generally accepted explanation to date.

In this work, we measured the relative orientation of spontaneously occurring RGBs in order to determine whether preferred orientations exist, and if so, we attempt to determine the factors that make them special in the hopes of obtaining useful information about their structure and related electronic properties.

Our results show that there are indeed preferred orientations with both small and large values of Σ, which, in agreement with the decades-old results of Chaudhari and Matthews, imply that the period of the interface is not a determining factor in them being preferred. Instead, special RGB orientations were characterized by a sharp decrease in the Shannon entropy [12] of their displacement spaces, implying a reduction in the number of possible Burgers vectors of interfacial dislocations.

2. Materials and Methods

Observations were made on a total of 58 RGBs obtained by chemical vapor deposition (CVD), as described in reference [13], exhibiting large areas of analyses. Rotation angles were measured directly from selected area electron diffraction (SAED) patterns obtained from individual RGBs and verified in high resolution transmission electron microscopy (HRTEM) using a transmission electron microscope, FEI-Titan, operated at 80 kV, to reduce the radiation damage to the samples.

3. Results

Figure 1a shows a typical SAED pattern and Figure 1b shows a high-resolution TEM image of two bi-layers with a 12.5° rotation angle. The distribution of rotation angles was found to be spread over the whole angular range of 0°–30°. Note that only this range needs to be considered, since the rotation angle θ and its complement $\theta_c = 60^\circ - \theta$ yield identical interfaces, except for a trivial 30° rigid body rotation. The results are summarized in Table 1 and the histogram shown in Figure 2. Additionally, we have measured the rotation angles in HRTEM images using the fast Fourier transform (FFT) in various images, as in the example shown in Figure 1b. In the table, the columns labeled θ and ξ, respectively, contain the angles measured as indicated in Figure 1 and a rational approximant ξ obtained from Equation (1). Since there are an infinite number of approximants for any given experimental accuracy, the ones given in Table 1 were calculated using continued fractions of increasingly higher orders until the difference $\Delta\theta$ (shown in the third column) between measured angles and those resulting from substituting the calculated approximant in Equation (1) was smaller than 0.05°. When the angular distance between two or more measured angles was smaller than the experimental error (0.3°), they were grouped together; in such cases, the value of θ that is given is the group's average.

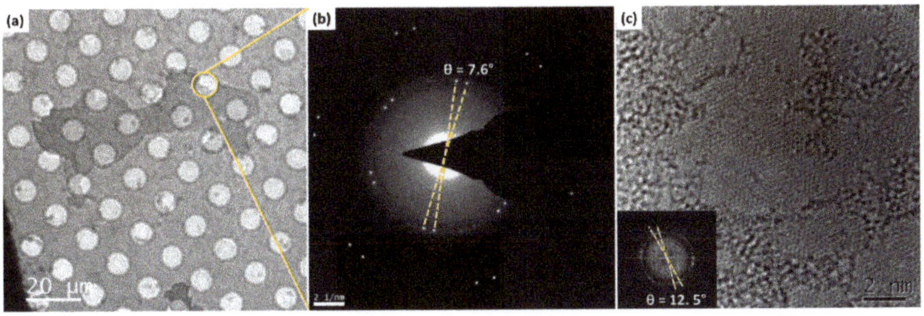

Figure 1. (**a**) Bright field image of the graphene sheets used to collect the SAED pattern. (**b**) SAED pattern of a bi-layer rotated by 7.6°. (**c**) High-resolution TEM image of two layers rotated 12.5°.

Table 1. List of measured rotation angles. The column labelled θ contains the measured angles in degrees. The column ξ contains the parameterized angles obtained from Equation (1) using continued fractions of increasingly higher orders until the difference $\Delta\theta$, shown in the third column between the rationally approximated angle and the measured one, was smaller than 0.05^o. An extra significant digit is given to the angles resulting from the average of two or more measurements.

$\theta(°)$	ξ	$\Delta\theta$	$\theta(°)$	ξ	$\Delta\theta$
5.70	33	0.03	18.30	$11 - 1/4$	0.01
6.80	29	0.04	19.03	$10 + 1/3$	0.00
7.60	$26 + 1/13$	0.00	19.70	10	0.00
9.70	$20 + 1/2$	0.04	20.00	$9 - 1/5$	0.05
10.23	$19 + 1/3$	0.01	20.60	$9 + 1/2$	0.03
10.63	$18 - 1/3$	0.03	21.50	$9 + 1/8$	0.00
11.50	$17 + 1/4$	0.03	22.80	$8 - 2/5$	0.03
12.53	$15 - 1/5$	0.01	23.50	$8 + 1/3$	0.02
13.40	$15 - 1/4$	0.01	23.90	$8 + 1/5$	0.05
14.15	14	0.04	24.50	$7 - 1/44$	0.02
14.90	$13 + 1/4$	0.00	24.80	$8 - 1/8$	0.01
16.03	$12 + 1/5$	0.05	26.80	$7 + 1/4$	0.04
16.50	$11 - 1/18$	0.00	27.70	$7 + 1/39$	0.02
17.20	$11 + 4/9$	0.01	30.00	$7 - 6/13$	0.01
17.80	$11 + 1/16$	0.00			

Figure 2 also shows the plot of the Shannon entropy [12] against θ of the displacement space, which is the space of all the possible Burgers vectors of interfacial dislocations [14]. The details of this calculation are given in the next section. For now, note that the entropy plot has sharp troughs at certain discrete values of θ, which are best described in terms of the parameterized rotation angle ξ, defined by

$$\theta = 2\tan^{-1}\frac{\sqrt{3}}{\xi} \tag{1}$$

Writing ξ as

$$\xi = x + \frac{k}{n} \tag{2}$$

With x; k and n integers, it becomes clear that entropy troughs are located at angles where k is either 1 or 0, as indicated by the value of ξ written below the most prominent troughs.

Note that all measured orientations closely match a sharp trough of the entropy curve where k is either 1 or 0. Even the angles $\theta = 17.2°, 22.8°$ for which $k \neq 0.1$, have decimal parts k/n very close to $1/2$; we have chosen not to make this approximation, however, in order to preserve the rule of keeping the difference between measured and rationally approximated angles smaller than $0.05°$. The other value with $k \neq 1$ at 30° is a secondary singular value, as explained below.The entropy plot was calculated with a an accuracy of 0.01°, which is an order of magnitude higher than the experimental error.

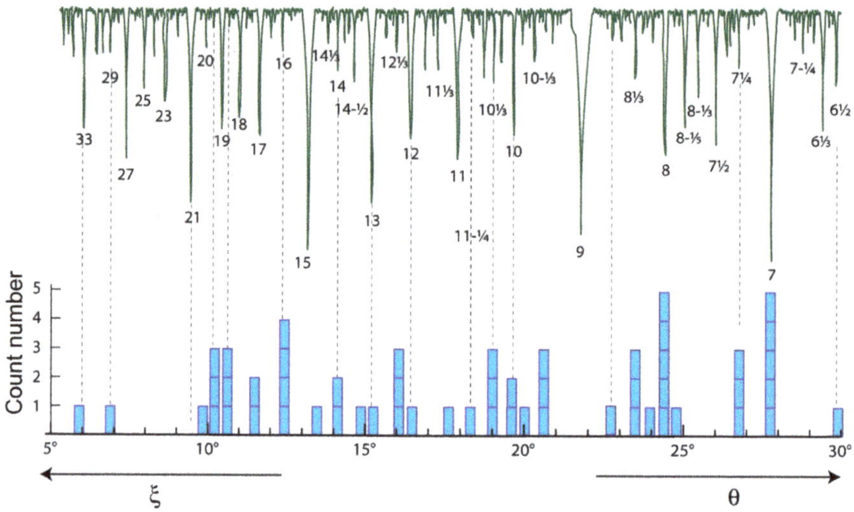

Figure 2. Bottom: Histogram of observed rotation angles between graphene sheets. The angular width of the bars is about 0.3° which is the estimated experimental error. Top: Plot of the Shannon entropy of the displacement (Burgers vector) space showing sharp troughs when $\xi = m$ or $m + 1/n$.

These results are also relatively close to the magic angles $> 5°$ reported by Tarnopolsky et. al., where their twisting angles followed a quasi-periodicity at $\Delta\alpha \simeq 3/2$ [9]. in our case, these matching angles presented sharp troughs in Figure 2, and also approximated the same periodicity. While most of the matching angles are reported in Table 1, the angle $\theta \sim 8.3°$ is included in the Shannon entropy calculation displayed in Figure 2. These angles are listed in Table 2.

Table 2. List of measured rotation angles and magic angles 5–11.5°.

	Matching Angles				
θ	5.7	6.8	8.3	9.7	11.5
α [9]	5.276	6.795	8.313	9.829	11.345

The physical explanation of these results can be simply stated in terms of the lattice type formed by the vectors of the displacement field of each interface, as discussed in the following subsections.

3.1. Primary and Secondary Singular States

According to Bollmann [14], there are special low energy interfaces he called "preferred states" which contain only primary dislocations, i.e., dislocations with Burgers vectors belonging to the crystal (graphene) lattices that compose the interface. Since Bollmann did not have a formal definition of a preferred state, he made the usual vague assumption that they were coincidence orientations with an unspecified small value of Σ. Bollmann further assumed that under small angular deviations, $\Delta\theta$ form a preferred state, and the interface would try to preserve its low energy configuration by concentrating the strain field in what he called secondary dislocations.In contrast with the primary ones, secondary dislocations have smaller non crystalline Burgers vectors belonging to the DSC lattice, which is the lattice formed by a set of vectors joining the sites of the composing lattices. Bollmann showed that secondary Burgers vectors could be calculated from a different interface formed by the DSC lattice, and a copy of itself rotated by $\Delta\theta$. We shall refer to this interface, which is central to the results of this work, as the "secondary interface" or "secondary state". The experimental observation of secondary dislocations [15], with dislocation spacings in accordance with Bollmann's hypotheses, confirmed the physical significance of the secondary interface. Preferred states will hereafter be referred to as singular states or singular interfaces to follow the more common usage [10].

It has been shown [16] that the rotation angles θ_x resulting from the substitution of an integer x for the rational parameter ξ in Equation (1) give rise to singular interfaces containing only primary dislocations. It is interesting to note that the set of integers actually partitions the angular space into equivalence classes $x - 1/2 \leqslant \xi < x + 1/2$ whose normal mode is the singular interface at θ_x [17].

In view of the above, any RGB with rotation angle $\theta = \theta_x + \Delta\theta$ can be thought of as having two components or states: a singular primary state with rotation angle θ_x and a secondary state with rotation angle $\Delta\theta$. It now makes sense to inquire under what circumstances the secondary state is, itself, singular.

To do this, we use Equation (1) to parameterize θ_x and $\Delta\theta$ as

$$\tan\frac{\theta_x}{2} = \frac{\sqrt{3}}{x}; \quad \tan\frac{\Delta\theta}{2} = \frac{\sqrt{3}}{\zeta} \tag{3}$$

for some of the integer x and rational ζ. After some trigonometry, we find that $\zeta = (3 + \xi x)/(x - \xi)$, and in substituting $x + \delta$ for ξ, we find that $-\zeta = x + \frac{T}{\delta}(3 + x^2)$, which shows that secondary states are, themselves, singular (integer ζ) when $\delta = 1/n$ or $\delta = (3 + x^2)/n$, i.e., when $k = 1$ in Equation (2) or $(3 + x^2)/n$ is an integer. This is the case of $\theta = 30^\circ$, since (see Table 1) $(3 + 7^2)/13$ is an integer. Therefore, we conclude that the minima in the entropy plot of Figure 2 correspond to RGBs which are either singular states or non-singular ones with a singular secondary component. In the next section, we describe the structural implications of this result.

3.2. The Displacement Space lattice

The formal framework for calculating the structure of the displacement space has been given elsewhere [16]. For completeness, we provide a simpler geometrical description here. Figure 3 shows, on the left, the dichromatic complex of a graphene bi-layer, and on the right, the set of (encircled) vectors join the displacement vectors, whose size is smaller than half a lattice vector of the graphene lattice. The displacement vectors that fulfill this condition are vectors of the DSC lattice that fall within a window given by the cut-projection formalism [16] (see Figure 4), and the circles around each vector are the locus of points of high strain and high electronic density in the interfacial plane between the graphene layers. This pattern not only provides a simpler crystallographic description of the interface by turning the dichromatic complex into a monochromatic one, but, more importantly, it establishes a mapping between the configuration and displacement spaces. Every vector in the displacement space is linked to all the sites of the monochromatic pattern that encircle the dichromatic points separated by that vector.

If ξ is a rational number, which it always is given the experimental error, then both the set of displacement vectors and its associated monochromatic pattern form a lattice whose degree of complexity depends on ξ. The complexity of the displacement space lattice (DSL) (multiplied by one inside the window and zero outside) is what the Shannon entropy calculation, shown in Figure 2, measures. Figure 4 shows examples of DSLs of varying complexities.

The entropy plot in Figure 2 was calculated by evaluating every 0.01° of the Shannon entropy of the distribution of bond lengths in the range [0.0–0.2] in lattice spacing units, divided into 30 intervals. Varying this range and/or the number of intervals may produce slight changes in the relative depths of the troughs, but there is always a sharp local entropy minimum when ξ is an integer (singular states) or when its decimal part is the inverse of an integer (secondary singular states).

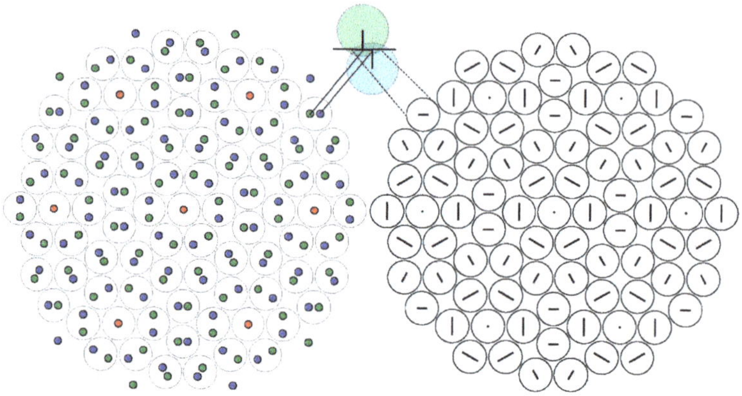

Figure 3. Left: dichromatic complex of the singular interface $\xi = 7, \Sigma = 13$ with green and blue circles representing the sites of each graphene sheet, as indicated in the central diagram illustrating the disposition of carbon atoms on either side of the interfacial plane. Red circles are coincidence sites. **Right**: Graphical representation of the displacement field as lines joining the sites, whose distance is smaller than half a lattice vector (one atomic radius). The circles around the midpoints of the displacement vectors simplify the crystallographic description of the interface by turning the dichromatic complex into a monochromatic one, whose sites are points of high strain where electron clouds overlap.

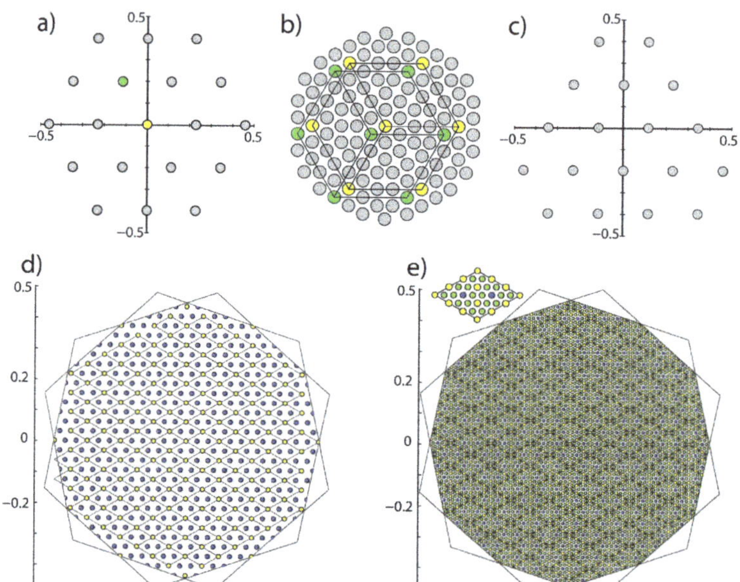

Figure 4. (**a**) Primitive DSL of the singular interface $\xi = 15, \Sigma = 19$ with both graphene sheets sharing a common origin. (**b**) Monochromatic pattern associated to DSL at left with green and yellow circles identifying sites with the displacements marked with the same colors in (**a**). (**c**) same DSL as (**a**) but with graphene sheets relatively displaced by the vector (0.12; 0.12) illustrating that relative sheet displacements only shift the DSL without altering its structure. (**d**,**e**) Non singular DSLs with primary states $\Sigma_{x=5} = 7$. (**d**) $\xi = 5 + \frac{1}{5}, \Sigma = 751$; (**e**) $\xi = 5 + \frac{3}{13}, \Sigma = 1591$ with an amplified unit cell at the top to make its contents clearer. The rotated hexagons identify the window used in the cut-projection formalism 14 used to calculate the DSLs.

4. Discussion

Given the hexagonal symmetry of the dichromatic pattern, each DSL point is associated with a subset of monochromatic sites that form a primitive hexagonal lattice (see Figure 4). An example of this is the hexagonal sublattice of coincidence sites associated with the DSL origin.

The DSL is always a simple primitive lattice when its unit cell is built using the basis vectors of the DSC lattice [18]. However, its degree of complexity is best described in terms of the subset of DSL points associated with hexagonal lattices, whose unit cells have the same areas as the unit cell of the CSL of the singular primary state; the physical significance of this unit cell is that its points represent Burgers vectors of primary dislocations. Using this unit cell, all primary states have a low entropy of the primitive DSL. The higher the order of the continued fraction expansion of ξ, the higher the contents of the unit cell. Figure 4 illustrates three levels of complexity of the displacement space. We conclude that crystallographic distinctions between DSLs, and their effects on the number of points that fit inside the window, accounts for the sharp entropy minima of singular and secondary singular interfaces.

In order to conclusively confirm that secondary singular interfaces are preferred states, ab-initio calculations need be performed in the vicinity of secondary singular interfaces at angular intervals close to the value of 0.01° used in the entropy calculation. This would entail calculations with much larger periodicities than the ones usually performed. We believe that the high computing cost of such calculations, coupled with the lack of a precise definition of singularity, have caused secondary singular interfaces to remain unnoticed.

In the meantime, although the Shannon entropy is not a measure of energy, it can be argued that since the latter depends on the distribution of interatomic distances, it is possible that interfaces with simpler distributions have lower energies. This is, after all, the role that dislocations play in reducing the elastic energy of crystalline materials.

Considering singular secondary interfaces as preferred (local energy minima) orientations not only accounts for the results obtained here for graphene, but it also explains other observations of spontaneously occurring nonsingular interfaces. For example, in the case of 001 twist interfaces in the cubic system, there are no singular orientations in the interval 36.875°, ($\xi = 3$) to 53.13°, ($\xi = 2$), both $\Sigma = 5$ interfaces. Yet, a number of interfaces in this range were observed by Chaudhari and Matthews [11]; in particular, those seen in their histogram near 39°; 41° and 42°, which appear to be the secondary singular $\xi = 3 - \frac{1}{6}(\theta = 38.88°, \Sigma = 325)$, $\xi = 3 - \frac{1}{3}(\theta = 41.11°, \Sigma = 73)$ and $\xi = 3 - \frac{1}{2}(\theta = 43.61°, \Sigma = 29)$. Although under the vague small Σ criterion, $\Sigma 29$ could have been considered a preferred state by Bollmann. The fact is that its DSL is not primitive, evidencing that a small value of Σ does not imply singularity.

5. Conclusions

Numerous calculations [10] have shown that singular interfaces have lower energies. This, however, does not answer the question of what happens with RGBs that start out far away from a singular orientation and do not have enough kinetic energy to rotate into a singular state. For example, Figure 2 shows that there are no singular RGBs between $\xi = 7(\theta = 27.8°, \Sigma = 13)$, and $\xi = 8(\theta = 24.4°, \Sigma = 67)$ so that any bi-layer starting to grow in the middle of this 3.4° interval would have to rotate more than one degree to find a singular orientation. Our results suggest that bi-layers starting out with a random rotation angle will try to rotate into a nearby secondary singular state. This appears to have been the case with the three $\Sigma = 889$ bi-layers observed at $\theta = 26.8°$.

The 0.3° experimental accuracy provided by the present experimental setup is not enough to experimentally confirm that secondary singular states are indeed the preferred low energy states; to do this, an experimental accuracy equal to, or better than, the 0.01° step used in the entropy calculation would be required. In spite of this, the consistent observation of RGBs near entropy minima, and the fact that secondary singular interfaces have been observed in other systems, supports this hypothesis.

Given the relevance of determining the mechanisms that control the precise orientation in designing materials with tailor-made properties, our results evidence the need to perform ab-initio calculations at, and close to, secondary singular orientations. Moreover, experiments using more accurate techniques to measure rotation angles are also needed to clarify this important point.

Author Contributions: Conceptualization and geometrical models, D.R.; Editing, electron microscopy analysis and measurements, A.P and A.H.-R.; Supervision A.P. All authors have read and agreed to the published version of the manuscript.

Funding: The authors thank the technical support in the use of the transmission electron microscope at CIQA. This work was supported by CONACYT FOMIX PUE-2018-03-02-84557 and A1-S-35309.

Institutional Review Board Statement: Not applicable.

Informed Consent Statement: Not applicable.

Data Availability Statement: Not applicable.

Acknowledgments: The authors wish to acknowledge the groups who provided the graphene samples: Jing Kong from the RLE-Nanomaterials and Electronics group at Massachusetts Institute of Technology MIT. A.H.R thanks to ConTex Doctoral Fellowship program #2019-000027-01EXTF-00125.

Conflicts of Interest: The authors declare no conflict of interest.

References

1. Zhang, Y.; Tang, T.T.; Girit, C.; Hao, Z.; Martin, M.C.; Zettl, A.; Crommie, M.F.; Shen, Y.R.; Wang, F. Direct observation of a widely tunable bandgap in bilayer graphene. *Nature* **2009**, *459*, 820–823. [CrossRef] [PubMed]
2. Radisavljevic, B.; Radenovic, A.; Brivio, J.; Giacometti, V.; Kis, A. Single-layer MoS2 transistors. *Nat. Nanotechnol.* **2011**, *6*, 147–150. [CrossRef] [PubMed]
3. Kam, K.K.; Parkinson, B.A. ChemInform Abstract: Detailed photocurrent spectroscopy of the semiconducting group vib transition metal dichalcogenides. *Chem. Inf.* **1982**, *13*, 463–467. [CrossRef]
4. Eda, G.; Yamaguchi, H.; Voiry, D.; Fujita, T.; Chen, M.; Chhowalla, M. Correction to Photoluminescence from Chemically Exfoliated MoS 2. *Nano Lett.* **2012**, *12*, 5111–5116. [CrossRef]
5. Mak, K.F.; Lee, C.; Hone, J.; Shan, J.; Heinz, T.F. Atomically thin MoS2: A new direct-gap semiconductor. *Phys. Rev. Lett.* **2010**, *105*, 136805–136809. [CrossRef]
6. De Trambly Laissardière, G.; Mayou, D.; Magaud, L. Localization of dirac electrons in rotated graphene bilayers. *Nano Lett.* **2010**, *10*, 804–808. [CrossRef] [PubMed]
7. Cao, Y.; Fatemi, V.; Demir, A.; Fang, S.; Tomarken, S.L.; Luo, J.Y.; Sanchez-Yamagishi, J.D.; Watanabe, K.; Taniguchi, T.; Kaxiras, E.; et al. Correlated insulator behaviour at half-filling in magic-angle graphene superlattices. *Nature* **2018**, *556*, 80–84. [CrossRef] [PubMed]
8. Cao, Y.; Fatemi, V.; Fang, S.; Watanabe, K.; Taniguchi, T.; Kaxiras, E.; Jarillo-Herrero, P. Unconventional superconductivity in magic-angle graphene superlattices. *Nature* **2018**, *556*, 43–50. [CrossRef] [PubMed]
9. Tarnopolsky, G.; Kruchkov, A.J.; Vishwanath, A. Origin of Magic Angles in Twisted Bilayer Graphene. *Phys. Rev. Lett.* **2019**, *122*, 106405. [CrossRef]
10. Priester, L. From Theory to Engineering; In *Grain Boundaries*; Springer: Dordrecht, The Netherlands, 2013; Volume 172.
11. Chaudhari, P.; Matthews, J.W. Coincidence twist boundaries between crystalline smoke particles. *J. Appl. Phys.* **1971**, *42*, 3063–3066. [CrossRef]
12. Shannon, C.E. A Mathematical Theory of Communication. *Bell Syst. Tech. J.* **1948**, *27*, 379–423. [CrossRef]
13. Reina, A.; Jia, X.; Ho, J.; Nezich, D.; Son, H.; Bulovic, V.; Dresselhaus, M.S.; Jing, K. Large area, few-layer graphene films on arbitrary substrates by chemical vapor deposition. *Nano Lett.* **2009**, *9*, 30–35. [CrossRef] [PubMed]
14. Bollmann, W. *Crystal Defects and Crystalline Interfaces*; Springer: Berlin/Heidelberg, Germany, 1970. [CrossRef]
15. Sun, C.P.; Balluffi, R.W. Secondary grain boundary dislocations in [001] twist boundaries in MgO I. Intrinsic structures. *Philos. Mag. A—Phys. Condens. Matter Struct. Defects Mech. Prop.* **1982**, *46*, 49–62. [CrossRef]
16. Romeu, D. Interfaces and quasicrystals as competing crystal lattices: Towards a crystallographic theory of interfaces. *Phys. Rev. B—Condens. Matter Mater. Phys.* **2003**, *67*, 024202–024234. [CrossRef]
17. Romeu, D.; Gómez-Rodríguez, A. Recurrence properties of O-lattices and the classification of grain boundaries. *Acta Crystallogr. Sect. A Found. Crystallogr.* **2006**, *62*, 411–412. [CrossRef] [PubMed]
18. Warrington, D.H. The coincidence site lattice (csl) and grain boundary (dsc) dislocations for the hexagonal lattice. *Le J. De Phys. Colloq.* **1975**, *36*, C4-87–C4-95. [CrossRef]

symmetry

MDPI

Article

Absolute Structure Determination of Chiral Zinc Tartrate MOFs by 3D Electron Diffraction

Christian Jandl [1], Gunther Steinfeld [1], Keyao Li [2], Pokka Ka Chuen Pang [2], Chun Lung Choi [2], Cengan Wang [2], Petra Simoncic [1,*] and Ian D. Williams [2,*]

1 Eldico Scientific AG, 5234 Villigen, Switzerland
2 Department of Chemistry, Hong Kong University of Science and Technology, Clear Water Bay, Kowloon, Hong Kong
* Correspondence: simoncic@eldico.ch (P.S.); chwill@ust.hk (I.D.W.); Tel.: +852-2358-7384 (I.D.W.)

Abstract: The absolute structure of the 3D MOF anhydrous zinc (II) tartrate with space group I222 has been determined for both [Zn(L-TAR)] and [Zn(D-TAR)] by electron diffraction using crystals of sub-micron dimensions. Dynamical refinement gives a strong difference in R factors for the correct and inverted structures. These anhydrous MOFs may be prepared phase pure from mild hydrothermal conditions. Powder X-ray diffraction indicates that isostructural or pseudo-isostructural phases can be similarly prepared for several other M^{2+} = Mg, Mn, Co, Ni and Cu. I222 is a relatively uncommon space group since it involves intersecting two-fold axes that place constraints on molecular crystals. However, in the case of MOFs the packing is dominated by satisfying the octahedral coordination centers. These MOFs are dense 3D networks with chiral octahedral metal centers that may be classed as Δ (for L-TAR) or Λ (for D-TAR).

Keywords: electron diffraction; MOFs; chirality; absolute structure

1. Introduction

The development of electron diffraction (ED, also termed 3D-ED or micro-ED) as a technique for structure determination applicable to organic molecular compounds, inorganics, and metal–organic frameworks (MOFs) has emerged in the last 10 years [1–7]. The advent of commercial electron diffractometers promises to usher in a new age of structural determination for crystalline materials that, either due to lack of material or inherent problems of crystal growth, cannot produce single specimens of suitable size for in-house X-ray diffractometers (typically around 50 μm, in certain cases even down to ca. 10 μm) [8].

In researching metal tartrates [9,10], which are of potential interest as fundamental chiral metal–organic frameworks that might have chiral separation [11], chiral catalytic [12], or chiro-optic applications [13] we found that simply formulated anhydrous materials such as [Zn(L-TAR)] could be prepared straightforwardly under mild hydrothermal conditions (Figure 1). At ambient conditions several metal tartrate phases had been crystallized and X-ray structures previously reported, with both coordinated aqua ligands and pore-included water molecules. These include [Zn(L-TAR)(H_2O)]·1.5H_2O (CSD codename CUJBAK) [14], [Zn(D-TAR)(H_2O)]·1.5H_2O (CUJBAK01) [15], a diastereomeric framework [Zn(L-TAR)(H_2O)]·2H_2O (KURNOB) [16], and a *meso* compound [Zn(m-TAR)$(H_2O)_2$] (MUYPUU) [17]. The structure of an anhydrous phase has not been reported in the Cambridge Structural Database (CSD) so far. This is likely due to the fact that whilst hydrothermal preparation of [Zn(L-TAR)] was possible, the resulting micro-crystalline powders precluded X-ray structure determination due to small particle size <10 microns.

The application of 3D-ED to the structure determination of MOFs with small particle size has been well established [5–7,18,19]. We were curious to see whether the as grown micro-crystals of [Zn(L-TAR)] could be amenable to structure determination using 3D-ED.

Furthermore, since the crystals should be homochiral, the determination of their absolute structure is also of interest. Recently it has been shown that by taking the effects of dynamical diffraction into account during refinement one can perform reliable absolute structure determinations from 3D-ED data of nanocrystalline samples [20,21].

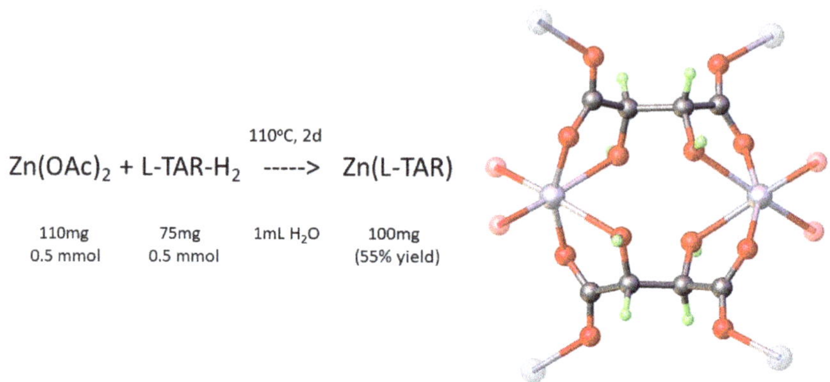

Figure 1. Synthesis and Structure of Molecular dimer SBU of product [Zn(L-TAR)] **1-L**. Structure diagram by Olex2 [22].

2. Materials and Methods

Metal acetates, tartaric acid and solvents used were of reagent grade supplied by Meryer Chemicals (Shanghai).

2.1. Preparation of Metal Tartrate Phases

[Zn(L-TAR)] **1-L**

This was prepared (Figure 1) by a one-pot hydrothermal reaction of zinc acetate hydrate, (0.5 mmol) and L-tartaric acid (0.5 mmol) in 1 mL water. Reagents were heated in a Teflon lined Parr pressure vessel (23 mL) for 2 d at 110 °C. Slow cooling afforded fine white microcrystalline powder. (Yield 55%)

[Zn(D-TAR)] **1-D** and other Zinc tartrates

The D-analogue was prepared in similar manner and yielded substituting D-tartaric acid. Use of racemic D/L-tartaric acid afforded a conglomerate of **1-L** and **1-D** based on the similar powder X-ray pattern obtained. Meso-tartaric acid afforded the known hydrated phase (coden MUYPUU [17]) up to 140 °C.

Other [M(L-TAR)] Phases

Other anhydrous divalent [M(L-TAR)] phases were prepared for M^{2+} = Mg, Mn, Co, Ni, and Cu in a similar manner to **1-L** at 110 °C/2d. These were shown to be isostructural, or pseudo-isostructural for Cu, by powder X-ray diffraction. The product phase for Ca^{2+} was different, identified as a higher coordinated hydrated phase. For Fe^{2+} an inhomogeneous mixture was obtained.

2.2. X-ray Crystallography

Powder X-ray diffraction data were obtained at room temperature using Cu-Kα radiation by a PanAlytical X'Pert PRO diffractometer with 1D X'celerator detector or on a PanAlytical Aeris benchtop powder X-ray diffractometer and measured in 2θ range 5 to 40° with step size of 0.02°.

2.3. Electron Crystallography

Samples were finely dispersed on standard TEM grids (amorphous carbon on Cu) and measured on an ELDICO *ED-1* electron diffractometer at room temperature using the software ELDIX [23]. The device is equipped with a LaB_6 source operating at an acceleration voltage of 160 kV (λ = 0.02851 Å) and a hybrid-pixel detector (Dectris QUADRO). Suitable crystals were identified in STEM imaging mode and diffraction was recorded in continuous rotation mode with a beam diameter of ca. 750 nm. The later parts of measurements showing significant beam damage were omitted. Further details can be found in Table 1.

Table 1. Three-dimensional-ED data collection details for the Zn tartrate samples (crystal size estimated from STEM images).

	[Zn(L-TAR)]	[Zn(D-TAR)]
angular range [°]	−60 to +80	−65 to +25
rotation per frame [°]	1.0	1.0
exposure time [s]	1.0	1.0
total exposure [s]	140	90
frames collected	140	90
frames used	1–80	1–70
crystal size [μm]	2.5 × 0.6 × 0.5	5.0 × 0.5 × 0.4

Data for kinematical refinement were processed and evaluated using the APEX4 software package [24]. After unit cell determination the frames were integrated and corrected for Lorentz effects, scan speed, background, and absorption using SAINT and SADABS [25,26]. Space group determination was based on systematic absences, E statistics, and successful refinement of the structure. The structure was solved using ShelXD and refined with ShelXL in conjunction with ShelXle [27–29]. Least squares refinements were carried out within the kinematic approximation by minimizing $\Sigma w(F_{obs}{}^2\text{-}F_{calc}{}^2)^2$ with the ShelXL weighting scheme and using neutral electron scattering factors [28,30]. Due to the low amount of data, all atoms were refined with isotropic displacement parameters. H atoms were placed in calculated positions based on typical distances for neutron diffraction and refined with a riding model and $U_{iso}(H) = 1.2{\cdot}U_{eq}(C)$, for hydroxy H atoms $U_{iso}(H) = 1.5{\cdot}U_{eq}(O)$ was used and angles were refined freely.

Data for dynamical refinement were processed using the PETS2 software package [31]. After unit cell determination the frames were integrated, corrected for pattern orientation and beam position, and merged into overlapping virtual frames with a tilt range of 3° and an offset between frames of 2° [21]. Dynamical refinement was performed using JANA2020 starting with the structure from kinematical refinement as initial model [32]. Least squares refinements were carried out by minimizing $\Sigma w(I_{obs} - I_{calc})^2$ based on dynamical diffraction intensities and assuming uniform thickness of the crystals. All atoms were refined with isotropic displacement parameters. Hydrogen atoms were treated in the same way as for kinematical refinement as a free refinement of bond lengths led to very long C-H bonds of ca. 1.3 Å, which seems excessive even though a certain elongation as compared to neutron diffraction data is expected [21].

2.4. *Differential Scanning Calorimetry/Thermal Gravimetric Analysis*

Thermal gravimetric analysis was conducted on TA instruments TGA analyzer under N_2 up to 800 °C. The plot (Figure 2, left) showed a single weight loss from [Zn(L-TAR)] (formula weight 213.5 g/mol) in the region 350–450 °C (−62%). Residual mass of 38% fits for ZnO (formula weight 81.4 g/mol). DSC measurements were made from ambient to 400 °C under a nitrogen atmosphere on a Universal V4.5A TA Instrument (Waters). A heating rate of 10 °C per minute was employed, a single endotherm was found with onset at 355 °C and with minimum heat flow at 375 °C, corresponding to the decomposition found in the TGA analysis.

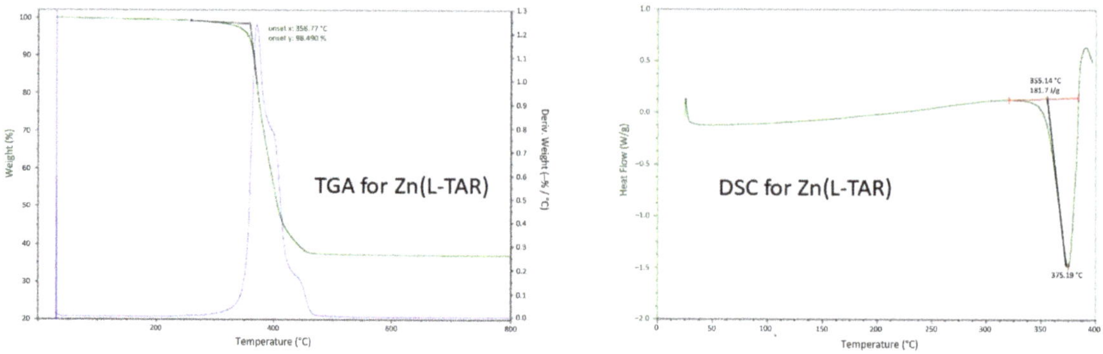

Figure 2. Thermal gravimetric analysis and differential scanning calorimetry for [Zn(L-TAR)] **1-L**.

3. Results

3.1. Hydrothermal Preparation of Zinc L-Tartrate and Related Phases

The structural data in the literature for zinc tartrates are dominated by hydrated phases that represent kinetic products of the system. In order to obtain the fundamental anhydrous zinc(II) tartrates, we employed mild hydrothermal synthesis. We found that conditions of 110 °C and 2 d were sufficient to eliminate crystallization of the hydrated phases. Notably, use of D/L-tartaric acid appeared to form a conglomerate with similar powder XRD to the [Zn(L-TAR)] and [Zn(D-TAR)] phases, albeit the pattern was a little broader (see Figure 3).

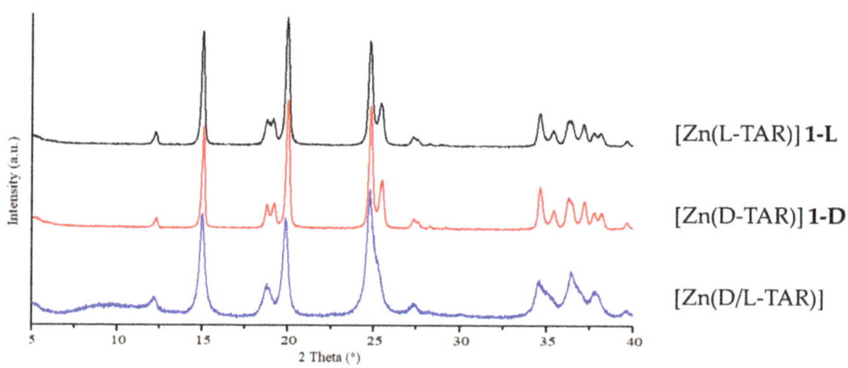

Figure 3. Powder X-ray diffractograms for [Zn(L-TAR)] **1-L**, [Zn(D-TAR)] **1-D**, and [Zn(D/L-TAR)].

3.2. Structural Analysis of Zinc Tartrate by Electron Diffraction

From the hydrothermal syntheses no crystals suitable for single crystal X-ray diffraction could be obtained as **1-L** and **1-D** only form very small needles which are also heavily intergrown. To circumvent these difficulties, we turned to 3D-ED, which allowed the structure determination directly from the as synthesized nanocrystalline samples.

1-L and **1-D** crystallize in the orthorhombic space group I222 with half a formula unit in the asymmetric unit (Tables 2 and 3 and Figures 1 and 4). They are isostructural to the reported Mn(II), Fe(II), Co(II), and Ni(II) tartrate coordination polymers, which are all based on naturally occurring L-tartrate [33]. Zn is coordinated by six O atoms in a distorted octahedral fashion with Zn-O distances between 2.00(2) Å and 2.36(4) Å. The longer Zn-O distances occur for the "backward" coordinating O of the carboxylate (see O1 in Figure 4) that chelates together with the hydroxy O forming a bite angle of 74.4(11)° to 75.8(7)° at the metal. The coordination sphere is completed by the same chelate motif of another tartrate molecule and two more carboxylate O atoms of different tartrates which coordinate in the common "forward" mode (see O2 in Figure 4). Thus, each Zn(II) ion is coordinated by four different tartrates and each tartrate is connected to four different Zn(II) ions forming a three-dimensional coordination polymer. The structure features no pores that could accommodate water or other solvent molecules and is essentially a 'condensed' phase.

Table 2. Crystal structure and refinement details for Zn(L-TAR) **1-L** and Zn(D-TAR) **1-D** (kin = kinematical refinement, dyn = dynamical refinement).

	1-L		**1-D**	
	kin	**dyn**	**kin**	**dyn**
CCDC number	2,242,890	2,242,892	2,242,891	2,242,894
chemical formula	$C_4H_4O_6Zn$			
formula weight [g mol^{-1}]	213.46			
crystal system	orthorhombic			
a [Å]	5.14(6)		5.16(6)	
b [Å]	8.97(10)		9.00(10)	
c [Å]	11.82(13)		11.78(13)	
α, β, γ [°]	90, 90, 90		90, 90, 90	
volume [Å^3]	545(11)		547(11)	
space group	I222			
Z	4			
resolution [Å]	0.79	0.68	0.79	0.74
completeness [%]	65	71	70	85
Unique reflections (measured/observed)	355/292	808/509	359/341	674/616
parameters	28	65	28	60
restraints	1	1	1	1
R_{int}	11.4	-	4.6	-
R_1(obs)/MR_1(obs) [1]	16.7	11.8/11.6	17.2	14.2/14.4
wR_2(all)	42.6	22.6	43.7	27.2
ΔV_{min}, ΔV_{max}	−0.29, 0.35	−0.16, 0.20	−0.32, 0.41	−0.22, 0.49

[1] MR: merged R factor for dynamical refinement based on post-refinement merging of reflections for better comparison with conventional R factor for kinematical refinement [21].

Table 3. Selected distances and angles for Zn(L-TAR) **1-L** and Zn(D-TAR) **1-D** (kin = kinematical refinement, dyn = dynamical refinement). Symmetry generated atoms marked by *.

	1-L		**1-D**	
	kin	**dyn**	**kin**	**dyn**
Zn-O1	2.02(3)	2.02(2)	2.02(3)	2.04(2)
Zn-O2 *	2.00(4)	2.03(2)	2.06(3)	2.00(2)
Zn-O3	2.29(4)	2.33(2)	2.36(4)	2.288(19)
C1-O1	1.24(3)	1.273(17)	1.24(4)	1.189(19)
C1-O2	1.29(3)	1.260(15)	1.21(3)	1.27(2)
C1-C2	1.51(3)	1.51(3)	1.59(4)	1.56(3)
C2-O3	1.37(3)	1.423(18)	1.31(4)	1.44(2)
C2-C2 *	1.55(4)	1.54(3)	1.52(5)	1.55(2)
O1-Zn1-O3	74.9(10)	75.8(7)	74.4(11)	74.5(7)
Zn1-O1-C1	120.5(15)	118.6(9)	121.3(18)	118.9(10)
Zn1-O3-C2	110.8(16)	110.1(9)	109(2)	112.8(8)
Zn1 *-O2-C1	130.4(18)	128.3(14)	125.2(19)	128.5(12)

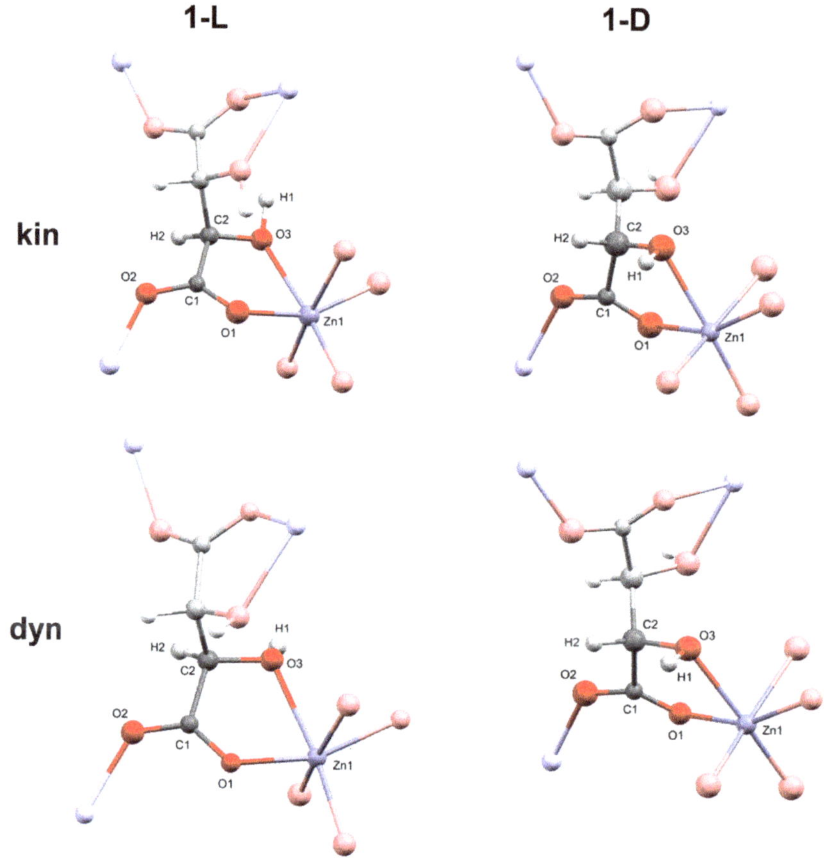

Figure 4. Expanded molecular structures for Zn(L-TAR) **1-L** and Zn(D-TAR) **1-D** (kin = kinematical refinement, dyn = dynamical refinement), symmetry equivalents are marked by pale print.

3.3. Absolute Structure Determination of Zinc Tartrates by Dynamical Refinement

As the title compound was synthesized with both enantiomers of tartaric acid, dynamical refinement of both enantiomorphs was performed for each sample to confirm the correct form. In both [Zn(L-TAR)] and [Zn(D-TAR)] the correct enantiomorph gives clearly better R/MR factors (MR_1 by 3.4% for **1-L** and 1.9% for **1-D**) than the inverted form (see Table 4). For further validation z-scores were calculated yielding values of 4.29σ for **1-L** and 3.50σ for **1-D** which corresponds to probabilities of >99,9% that the absolute structures are determined correctly (see Table 4) [21]. In addition, of course, the R factors also improve in comparison to the kinematical refinement results (see Table 2).

Table 4. Dynamical refinement results for enantiomorphic Zn(L-TAR) **1-L** and Zn(D-TAR) **1-D**.

Sample	1-L		1-D	
enantiomorph	L-form	D-form	L-form	D-form
R_1(obs)/MR_1(obs)	11.8/11.6	14.5/15.0	16.1/16.3	14.2/14.4
wR_2(all)	22.6	28.4	30.5	27.2
z-score [1]	4.288σ		3.502σ	
probability [1]	99.999%		99.977%	

[1] z-score and corresponding probability for correct enantiomorph determination based on the observed statistics of reflections with better fit for the correct enantiomorph compared to a normal distribution of random differences [21].

Notably the H position on the hydroxy group differs between the dynamically (dyn) and kinematically (kin) refined structures of **1-L** while it is similar for **1-D**. The H position of **1-L**-dyn corresponds to a hydrogen bond with a carboxylate O atom from an adjacent ZO_6 octahedron at a O-O distance of 3.21(3) Å. It also agrees with the literature reported structure of the Co(II) analogue [33]. The H position of **1-L**-kin seems rather unusual, it features an unnatural C-O-H angle of 98(6)° and leads to a contact of 2.1(2) Å with its symmetry equivalent. Although its position would at least correspond to a potential hydrogen bond to a carboxylate O of the same tartarte molecule at a distance of 3.39(5) Å, one cannot rely on the H position from kinematical refinement in this case. Starting the refinement from the correct position obtained from **1-L**-dyn still produces the same result.

3.4. *Preparation and Characterization of Related Metal Tartrate Phases*

As mentioned the structure of **1-L** and **1-D** was found analogous to several other anhydrous M(II) tartrates, which were able to be characterized by single crystal X-ray diffraction [33–35]. Similar hydrothermal syntheses (110 °C, 2 d, 0.5 mmol scale in 1 mL water) using various divalent metals M^{2+} = Mg, Mn, Co, Ni, and Cu were carried out and the micro-crystalline powders analyzed by powder X-ray diffraction. The results are shown in Figures 5 and 6.

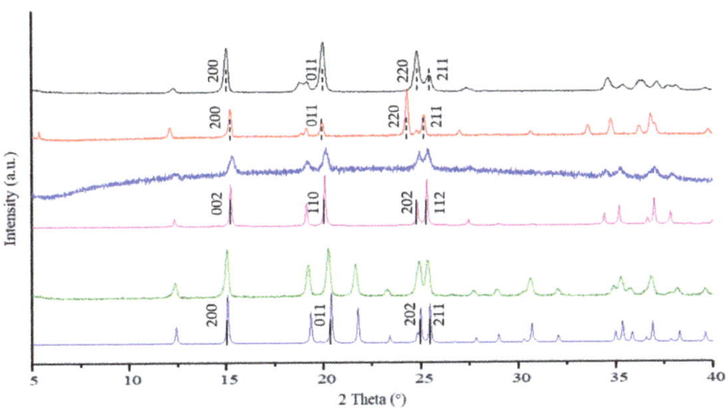

Figure 5. Powder X-ray diffractograms for **1-L** and Mn, Co, and Cu metal tartrates [M(L-TAR)].

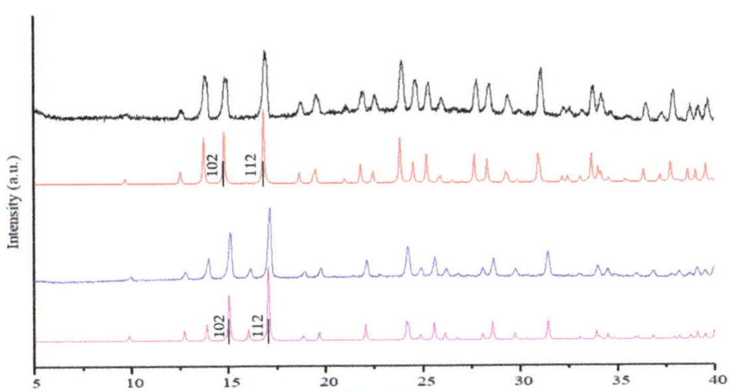

Figure 6. Powder X-ray diffractograms for Mg and Ni metal tartrates [M(L-TAR)].

The powder X-ray diffractograms of anhydrous [Co(L-TAR)] and [Cu(L-TAR)] are calculated from the single crystal structures of ACOVEU [33] and VIJGUS [36], respectively.

The solid lines represent the actual crystal planes from the single crystal structures, while the dashed lines show the hkl indexed by DICVOL06 [37]. The indexed planes of the main peaks all fit with the equivalent planes of the actual cells, showing Zn, Mn, Co, and Cu formed anhydrous metal tartrates using these hydrothermal conditions.

The powder X-ray diffractograms of [Mg(D-TAR)(H_2O)·1.5H_2O] and [Ni(L-TAR)(H_2O)·1.5H_2O] hydrates are calculated from the single crystal structures of JIFXIG [34] and CIXKEZ02 [35], respectively. The two Mg and Ni metal tartrate products match well with their corresponding calculated patterns, indicating that these two metals tend to form hydrates under 110 °C hydrothermal conditions.

4. Discussion

4.1. Structure and Absolute Structure Determination

The zinc tartrate phases represent a typical application example for the 3D-ED technique when X-ray crystallography fails due to small particle size < 1 μm. The diffraction patterns from numerous specimens were scanned but the majority showed twinning or intergrowth problems. One benefit of electron diffraction is that in imaging mode individual crystals can quickly be identified and then studied for diffraction. After a suitable crystal was identified by its pattern, the intensity data were collected in continuous rotation mode. Due to beam damage the maximum exposure for crystals is limited before structural degradation occurs, so a rapid intensity data collection was necessitated. The orthorhombic symmetry of the pattern meant that *hkl* coverage was acceptable just from one crystallite.

The successful solution and reasonable quality of the refined structural model using the kinematic approximation gave R_1 = 16.7% for [Zn(L-TAR)]. This then led us to attempt a dynamical refinement in order to determine the absolute structure. This was carried out using JANA2020 [32] which is so far the only publicly available program with this ability and has been used to establish absolute structures in various cases, most of them organic molecules related to pharma [20,21]. The application of this approach gave R_1 = 11.8% (MR_1 = 11.6%) for the correct hand (2R, 3R) and R_1 = 14.5% (MR_1 = 15.0%) for the inverted (2S, 3S) structure, whereas there is no difference in R_1 at all for the inversion of the kinematically refined structure. The difference of approximately 3 percentage points along with a z-score of ca. 4.3σ allows for unequivocal absolute structure identification. As a confirming experiment, a fresh sample of [Zn(D-TAR)] was prepared and analyzed in the same way. The results on the **1-D** specimen are in good agreement with the data for **1-L** except that the dynamical diffraction models indicate the inverse absolute stereochemistry. The data are summarized in the above tables.

So dynamical refinement indeed proved to be a reliable method for the identification of the stereochemistry in these nanocrystalline samples. Dynamical refinement also proved to be superior in the location of H atoms as the kinematical refinement of **1-L** yielded a wrong H position on the hydroxy group. Even though there have been reports of H atom locations based on just kinematical refinement results, care has to be taken in the evaluation of such results and often dynamical refinement will be necessary for the accurate identification of all H sites [21]. As nanocrystalline samples such as in our case are a common challenge encountered in MOF chemistry, 3D-ED has already gained considerable attention in this field [2,4–7]. The ability to determine crystal structures from just a single crystallite and even identify the absolute structure in the cases of chiral compounds will likely let the use of 3D-ED grow rapidly in the future.

4.2. Topological Comparison of Zinc Tartrate Phases

The geometric details for the structure (metal–ligand and C-O bond lengths) are in general agreement with expectations (Table 3), but the bond length uncertainties are unavoidably higher for 3D-ED than for SC-XRD. The anhydrous phase can be briefly compared with the previously reported zinc tartrates reported in the Cambridge Structural Database [14–17].

The next level of structural comparison for these MOFs is the structural building block SBU. This is best seen as a molecular dimer comprising two octahedral M^{2+} dications and two bridging μ^2-tartrate-dianion ligands. In the anhydrous phase both tartrates are μ^4 and provide six coordination sites for the metal. In the hydrated phases CUJBAK and KURNOB the two tartrates are chemically distinct with one μ^4- and one μ^2-tartrate. The keto oxygens of one ligand are not coordinated and the two vacated sites are occupied by coordinated aqua groups. The distinction between the two hydrated frameworks is that they are diastereomeric—the water and keto coordination are reversed as may be seen in Figure 7. It may also be pointed out that the octahedral metal centers are also chiral and the designation of Δ can be ascribed for L-tartrate networks and Λ for D-tartrate analogues.

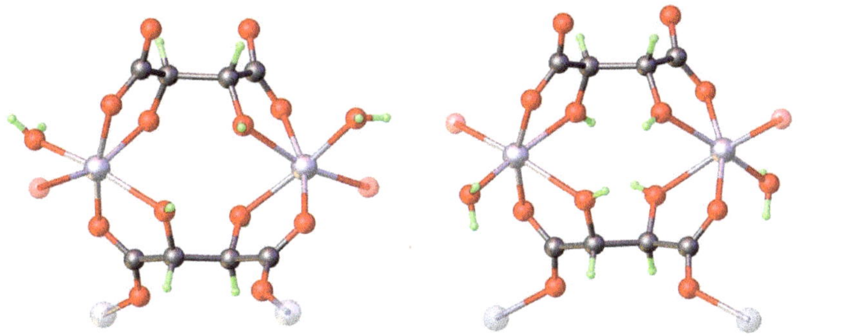

Figure 7. Metal dimer SBU in hydrated metal L-tartrate MOFs (CUJBAK **left**) and KURNOB **right**).

Finally, the overall assembly of the [M(TAR)] SBUs into the MOF framework can be considered. The I222 phase is completely 3D, whereas CUJBAK and KURNOB form 2D topological networks (see Figure 8) that are distinct from each other and also entrap different amounts of guest water molecules. The arrangement for I222 amounts to combining the connectivity found in the two 2D networks since the dimer unit is the same, but the keto ligation is the combination of what is found in the other two hydrated phases (see Figure 9).

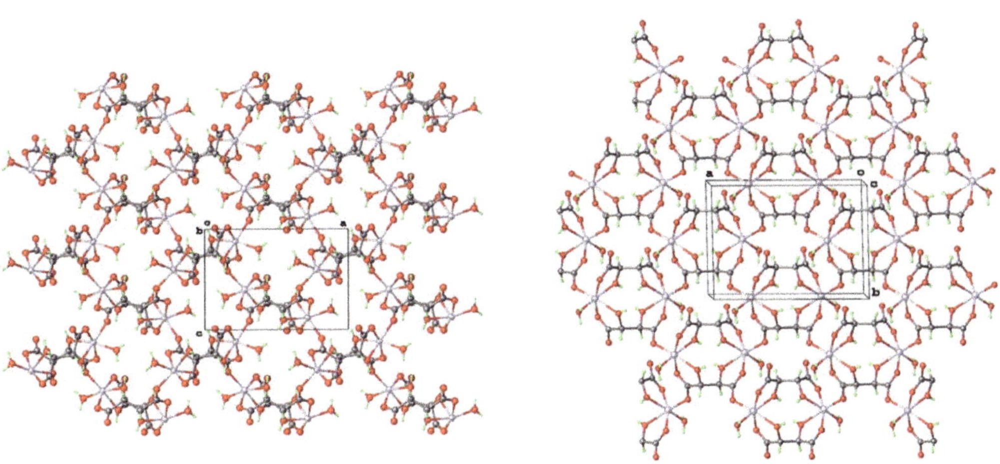

Figure 8. Two-dimensional sheet frameworks for CUJBAK **left** (viewed along [010]) and KURNOB **right** (viewed along [001]). Unit cell axes and origin labelled a, b, c and o.

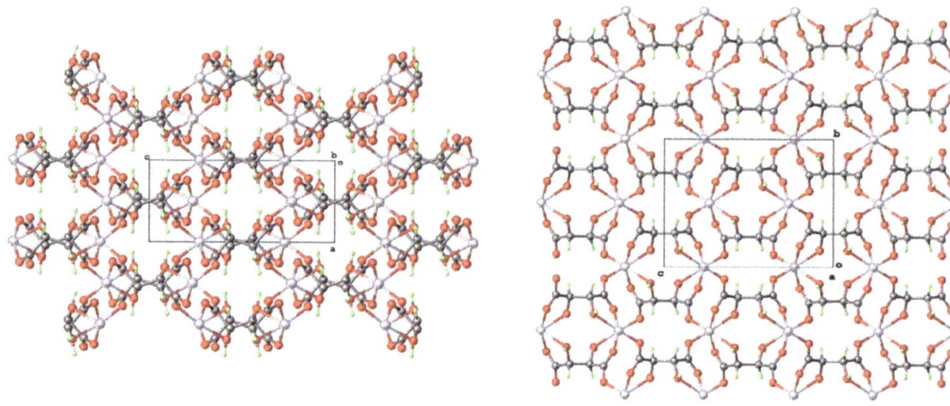

Figure 9. Packing diagram for **1-L** (from dynamical refinement) viewed along [010] **left** and [100], **right**.

The meso phase MUYPUU has two coordinated aqua molecules per Zn and this forms a 1D chain polymer, that packs in an efficient interdigitated manner with itself, entrapping no further water of crystallization (see Figure 10). This may hint at why this hydrated form persists to even higher temperatures than the L-tartrate hydrated phases, since hydrothermal preparation at 140 °C still yielded this hydrated phase type.

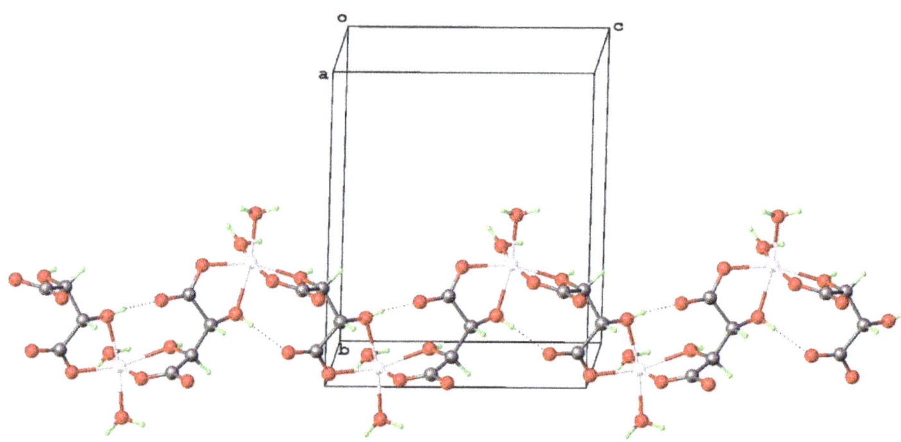

Figure 10. Packing diagram for MUYPUU [Zn(m-TAR)(H_2O)$_2$] viewed close to [100].

4.3. Isostructurality with Other Bivalent Metal Tartrates

Hydrothermal preparation of [M(L-TAR)] analogues with Zn were carried out under similar conditions for M^{2+} = Mg, Ca, Mn, Fe, Co, Ni, and Cu. In all cases microcrystalline powders resulted. Powder X-ray diffractograms indicated that isostructural or pseudo isostructural phases could be prepared for most of these metals with the exception of Ca and Fe. The preparation of [Fe(L-TAR)] using Fe(OAc)$_2$ in a similar manner was unsuccessful, although [M(L-TAR)] phases M = Mn, Ni, Co, and Fe have been prepared and investigated for their magnetic properties [33]. The structures of [Ni(L-TAR)] [35] and [Cu(L-TAR)] [36] were obtained after dehydration of hydrated phases through single-crystal to single-crystal transformations. Calcium prefers to adopt a hydrated structure with higher coordination number [38].

Reactions with L-(2R,3R)-tartaric and D-(2S,3S)-tartaric acid afforded the mirror image chiral MOFs **1-L** and **1-D**. Notably tartaric acid has two other forms *meso*-2R,3S-tartaric acid and racemic D/L-(2R,3R/2S,3S)-tartaric acid. Corresponding reactions substituting these different stereochemical forms could allow for phase types with different crystal structures incorporating inversion symmetry. However, the Zn(D/L-TAR) reaction gives a powder XRD pattern that seems close to the L-TAR or D-TAR products indicating that a conglomerate of these crystals is favored. In the *meso*-case the centrosymmetric hydrated phase MUYPUU is still afforded at 110 °C. This phase also persists at 140 °C, but higher temperature may still result in an anhydrous phase. Table 5 gives the unit cell data for the phase types and corresponding entries that were retrieved from the CSD or PDF for the bivalent anhydrous metal tartrates mentioned above.

Table 5. Crystal unit cell data for related anhydrous M(TAR) crystal phases.

	Mg(D-TAR)	Co(L-TAR)	Ni(L-TAR)	Cu(L-TAR)
reference	[34]	[33]	[35]	[36]
CCDC number	631943	230223	650895	927769
empirical formula	$C_4H_4O_6Mg$	$C_4H_4O_6Co$	$C_4H_4O_6Ni$	$C_4H_4O_6Cu$
formula weight	172.38	207	207.78	211.61
temperature [K]	293	120(2)	113(2)	200(2)
crystal system	orthorhombic	orthorhombic	orthorhombic	orthorhombic
space group	I222	I222	I222	$P2_12_12$
a [Å]	5.024(1)	5.057(2)	4.9740(11)	4.9808(5)
b [Å]	9.163(2)	9.117(3)	9.055(2)	8.9689(11)
c [Å]	11.455(2)	11.631(3)	11.472(2)	11.7511(13)
α [°]	90	90	90	90
β [°]	90	90	90	90
γ [°]	90	90	90	90
volume [Å^3]	527.2(2)	536.2(3)	516.68(19)	524.95(10)

Investigations of structure property relations for various optical and electrooptical effects are in progress including chiro-photoluminescence of zinc tartrate.

5. Conclusions

The crystal structures of the 3D network coordination polymers [Zn(L-TAR)] and [Zn(D-TAR) have been investigated by 3D-ED and were solved and refined using conventional X-ray crystallography programs with both a kinematical and dynamical diffraction approach. The absolute structures could also clearly be discriminated. Symmetry analysis of the network in these 3D nets indicate that not only is the ligand chiral with two halves related by a two-fold axis, but the octahedral metal centers are chiral as well and can be classified as Δ or Λ configuration. The current work shows that valuable and detailed structural information on chiral nanocrystalline MOF materials can be provided by modern electron diffraction techniques.

Author Contributions: Conceptualization, I.D.W.; synthesis and methodology, P.K.C.P. and C.L.C.; validation, K.L. and C.W.; formal analysis, I.D.W. and C.J.; investigation, K.L., C.J. and G.S.; data curation, C.J.; writing—original draft preparation, I.D.W. and C.J.; writing—review and editing, All; supervision, I.D.W.; project administration, I.D.W. and P.S.; funding acquisition, I.D.W. All authors have read and agreed to the published version of the manuscript.

Funding: Funding support of the Research Grants Council of Hong Kong (C6022-20E) and the Hong Kong Branch of the Southern Marine Science and Engineering Guangdong Laboratory (Guangzhou) (grant No. SMSEGL-20SC01-D) are gratefully acknowledged.

Data Availability Statement: CCDC 2242890-2242895 contain the supplementary crystallographic data for this paper. These data are provided free of charge by the joint Cambridge Crystallographic Data Centre and Fachinformationszentrum Karlsruhe Access Structures Service www.ccdc.cam.ac.uk/structures (accessed on 22 February 2023).

Acknowledgments: We thank Jane Y C. Wu for help with the DSC measurements.

Conflicts of Interest: The authors declare no conflict of interest. The funders had no role in the design of the study; in the collection, analyses, or interpretation of data; in the writing of the manuscript; or in the decision to publish the results.

References

1. Nannenga, B.L.; Gonen, T. The Cryo-EM Method Microcrystal Electron Diffraction (MicroED). *Nat. Methods* **2019**, *16*, 369–379. [CrossRef] [PubMed]
2. Gruene, T.; Wennmacher, J.T.C.; Zaubitzer, C.; Holstein, J.J.; Heidler, J.; Fecteau-Lefebvre, A.; De Carlo, S.; Müller, E.; Goldie, K.N.; Regeni, I.; et al. Rapid Structure Determination of Microcrystalline Molecular Compounds Using Electron Diffraction. *Angew. Chem. Int. Ed.* **2018**, *57*, 16313–16317. [CrossRef] [PubMed]
3. Jones, C.G.; Martynowycz, M.W.; Hattne, J.; Fulton, T.J.; Stoltz, B.M.; Rodriguez, J.A.; Nelson, H.M.; Gonen, T. The CryoEM Method MicroED as a Powerful Tool for Small Molecule Structure Determination. *ACS Cent. Sci.* **2018**, *4*, 1587–1592. [CrossRef] [PubMed]
4. Gruene, T.; Holstein, J.J.; Clever, G.H.; Keppler, B. Establishing Electron Diffraction in Chemical Crystallography. *Nat. Rev. Chem.* **2021**, *5*, 660–668. [CrossRef]
5. Huang, Z.; Grape, E.S.; Li, J.; Inge, A.K.; Zou, X. 3D Electron Diffraction as an Important Technique for Structure Elucidation of Metal-Organic Frameworks and Covalent Organic Frameworks. *Coord. Chem. Rev.* **2021**, *427*, 213583. [CrossRef]
6. Huang, Z.; Willhammar, T.; Zou, X. Three-Dimensional Electron Diffraction for Porous Crystalline Materials: Structural Determination and Beyond. *Chem. Sci.* **2021**, *12*, 1206–1219. [CrossRef]
7. Samperisi, L.; Jaworski, A.; Kaur, G.; Lillerud, K.P.; Zou, X.; Huang, Z. Probing Molecular Motions in Metal–Organic Frameworks by Three-Dimensional Electron Diffraction. *J. Am. Chem. Soc.* **2021**, *143*, 17947–17952. [CrossRef]
8. Hovestreydt, E. Electron Diffraction: Accelerating Drug Development. *Drug Discov. Today* **2022**, *27*, 371–373. [CrossRef]
9. Au-Yeung, A.S.-F.; Sung, H.H.-Y.; Cha, J.A.K.; Siu, A.W.-H.; Chui, S.S.-Y.; Williams, I.D. Hydrothermal Synthesis of Indium Tartrates: Structures of the Chiral Polymer [In(L-TAR)$^{3-}$$H_2O$]·0.5$H_2O$ Containing the Tartrate Trianion, and a Microporous Hybrid Solid [In(OH)(D/L-TAR)$^{2-}$]·2H_2O. *Inorg. Chem. Commun.* **2006**, *9*, 507–511. [CrossRef]
10. Thushari, S.; Cha, J.A.K.; Sung, H.H.-Y.; Chui, S.S.-Y.; Leung, A.L.-F.; Yen, Y.-F.; Williams, I.D. Microporous Chiral Metal Coordination Polymers: Hydrothermal Synthesis, Channel Engineering and Stability of Lanthanide Tartrates. *Chem. Commun.* **2005**, 5515–5517. [CrossRef]
11. Lu, Y.; Zhang, H.; Chan, J.Y.; Ou, R.; Zhu, H.; Forsyth, M.; Marijanovic, E.M.; Doherty, C.M.; Marriott, P.J.; Holl, M.M.B.; et al. Homochiral MOF–Polymer Mixed Matrix Membranes for Efficient Separation of Chiral Molecules. *Angew. Chem. Int. Ed.* **2019**, *58*, 16928–16935. [CrossRef]
12. Yoon, M.; Srirambalaji, R.; Kim, K. Homochiral Metal–Organic Frameworks for Asymmetric Heterogeneous Catalysis. *Chem. Rev.* **2012**, *112*, 1196–1231. [CrossRef]
13. Hu, L.; Li, K.; Shang, W.; Zhu, X.; Liu, M. Emerging Cubic Chirality in γCD-MOF for Fabricating Circularly Polarized Luminescent Crystalline Materials and the Size Effect. *Angew. Chem. Int. Ed.* **2020**, *59*, 4953–4958. [CrossRef] [PubMed]
14. Templeton, L.K.; Templeton, D.H.; Zhang, D.; Zalkin, A. Structure of Di-μ-(+)-Tartrato-Bis[Aquazinc(II)] Trihydrate, [$Zn_2(C_4H_4O_6)_2(H_2O)_2$]·3$H_2O$ and Anomalous Scattering by Zinc. *Acta Crystallogr. Sect. C* **1985**, *41*, 363–365. [CrossRef]
15. Lin, H.-Y.; Hu, H.-L.; Chen, B.-K.; Li, J. Synthesis and Characterization of a New Two-Dimensional Compound: {[$Zn(C_4H_4O_6)(H_2O)$] 23H_2O}$_n$. *Chin. J. Spectrosc. Lab.* **2009**, *26*, 803–806.
16. Liu, H.-T.; Lu, J.; Wang, D.-Q. Poly[[Diaqua(μ_4-L-Tartrato)(μ_2-L-Tartrato)Dizinc(II)] Tetrahydrate]. *Acta Crystallogr. Sect. E* **2010**, *66*, m374. [CrossRef]
17. Li, M.-C.; Zhang, J.-R.; Yuan, S.-X.; Zhang, D. Crystal Structure of Catena-Poly[Diaqua-(μ_2-Tartrato-κ^4O,O′:O″,O‴)Zinc(II)], $C_4H_8O_8Zn$. *Z. Kristallogr.-N. Cryst. Struct.* **2020**, *235*, 1221–1222. [CrossRef]
18. Hynek, J.; Brázda, P.; Rohlíček, J.; Londesborough, M.G.S.; Demel, J. Phosphinic Acid Based Linkers: Building Blocks in Metal–Organic Framework Chemistry. *Angew. Chem. Int. Ed.* **2018**, *57*, 5016–5019. [CrossRef]
19. Portolés-Gil, N.; Lanza, A.; Aliaga-Alcalde, N.; Ayllón, J.A.; Gemmi, M.; Mugnaioli, E.; López-Periago, A.M.; Domingo, C. Crystalline Curcumin BioMOF Obtained by Precipitation in Supercritical CO_2 and Structural Determination by Electron Diffraction Tomography. *ACS Sustain. Chem. Eng.* **2018**, *6*, 12309–12319. [CrossRef]
20. Brázda, P.; Palatinus, L.; Babor, M. Electron Diffraction Determines Molecular Absolute Configuration in a Pharmaceutical Nanocrystal. *Science* **2019**, *364*, 667–669. [CrossRef]
21. Klar, P.; Krysiak, Y.; Xu, H.; Steciuk, G.; Cho, J.; Zou, X.; Palatinus, L. Chirality and Accurate Structure Models by Exploiting Dynamical Effects in Continuous-Rotation 3D ED Data. *ChemRxiv* **2021**. [CrossRef]
22. Dolomanov, O.V.; Bourhis, L.J.; Gildea, R.J.; Howard, J.a.K.; Puschmann, H. OLEX2: A Complete Structure Solution, Refinement and Analysis Program. *J. Appl. Cryst.* **2009**, *42*, 339–341. [CrossRef]
23. *ELDIX Software Suite*, Version 0.14.0; ELDICO Scientific AG: Villigen, Switzerland, 2022.
24. *Apex Suite of Crystallographic Software*, APEX4, Version 2022.1-1; Bruker AXS Inc.: Madison, WI, USA, 2022.
25. *SAINT*, Version 8.40B; Bruker AXS Inc.: Madison, WI, USA, 2019.

26. *SADABS*, Version 2016/2; Bruker AXS Inc.: Madison, WI, USA, 2016.
27. Sheldrick, G.M. A Short History of SHELX. *Acta Crystallogr. Sect. A* **2008**, *64*, 112–122. [CrossRef]
28. Sheldrick, G.M. Crystal Structure Refinement with SHELXL. *Acta Crystallogr. Sect. C* **2015**, *71*, 3–8. [CrossRef]
29. Hübschle, C.B.; Sheldrick, G.M.; Dittrich, B. ShelXle: A Qt Graphical User Interface for SHELXL. *J. Appl. Cryst.* **2011**, *44*, 1281–1284. [CrossRef]
30. Doyle, P.A.; Turner, P.S. Relativistic Hartree–Fock X-Ray and Electron Scattering Factors. *Acta Crystallogr. Sect. A* **1968**, *24*, 390–397. [CrossRef]
31. Palatinus, L.; Brázda, P.; Jelínek, M.; Hrdá, J.; Steciuk, G.; Klementová, M. Specifics of the Data Processing of Precession Electron Diffraction Tomography Data and Their Implementation in the Program PETS2.0. *Acta Crystallogr. Sect. B* **2019**, *75*, 512–522. [CrossRef]
32. Petříček, V.; Dušek, M.; Palatinus, L. Crystallographic Computing System JANA2006: General features. *Z. Kristallogr.-Cryst. Mater.* **2014**, *229*, 345–352. [CrossRef]
33. Coronado, E.; Galán-Mascarós, J.R.; Gómez-García, C.J.; Murcia-Martínez, A. Chiral Molecular Magnets: Synthesis, Structure, and Magnetic Behavior of the Series [M(L-Tart)] (M = Mn^{II}, Fe^{II}, Co^{II}, Ni^{II}; L-Tart = (2R,3R)-(+)-Tartrate). *Chem.–Eur. J.* **2006**, *12*, 3484–3492. [CrossRef]
34. Kam, K.C.; Young, K.L.M.; Cheetham, A.K. Chemical and Structural Diversity in Chiral Magnesium Tartrates and Their Racemic and Meso Analogues. *Cryst. Growth Des.* **2007**, *7*, 1522–1532. [CrossRef]
35. Zhu, P.; Gu, W.; Zhang, L.-Z.; Liu, X.; Tian, J.-L.; Yan, S.-P. A Rare Thermally Induced Single Crystal to Single Crystal Transformation from a 2D Chiral Coordination Polymer to a 3D Chiral Coordination Polymer. *Eur. J. Inorg. Chem.* **2008**, *2008*, 2971–2974. [CrossRef]
36. Liu, Y.-H.; Lee, S.-H.; Chiang, J.-C.; Chen, P.-C.; Chien, P.-H.; Yang, C.-I. Dehydration Induced 2D-to-3D Crystal-to-Crystal Network Re-Assembly and Ferromagnetism Tuning within Two Chiral Copper(II)–Tartrate Coordination Polymers. *Dalton Trans.* **2013**, *42*, 16857–16867. [CrossRef] [PubMed]
37. Boultif, A.; Louer, D. Powder Pattern Indexing with the Dichotomy Method. *J. Appl. Cryst.* **2004**, *37*, 724–731. [CrossRef]
38. Appelhans, L.N.; Kosa, M.; Radha, A.V.; Simoncic, P.; Navrotsky, A.; Parrinello, M.; Cheetham, A.K. Phase Selection and Energetics in Chiral Alkaline Earth Tartrates and Their Racemic and Meso Analogues: Synthetic, Structural, Computational, and Calorimetric Studies. *J. Am. Chem. Soc.* **2009**, *131*, 15375–15386. [CrossRef] [PubMed]

symmetry

MDPI

Article

Influence of Precession Electron Diffraction Parameters and Energy Filtering on Reduced Density Function Analysis of Thin Amorphous Silica Films—Implications for Structural Studies

Yu-Jen Chou [1,*], Konstantin B. Borisenko [2,*], Partha Pratim Das [3], Stavros Nicolopoulos [3], Mauro Gemmi [4] and Angus I. Kirkland [5]

1 Department of Mechanical Engineering, National Taiwan University of Science and Technology, No. 43, Sec. 4, Keelung Road, Taipei 10607, Taiwan
2 The Kennedy Institute of Rheumatology, University of Oxford, Oxford OX3 7FY, UK
3 NanoMEGAS SPRL, Rue Èmile Claus 49 bte 9, 1050 Brussels, Belgium; partha@nanomegas.com (P.P.D.); info@nanomegas.com (S.N.)
4 Istituto Italiano di Tecnologia, Center for Materials Interfaces, Electron Crystallography, Viale Rinaldo Piaggio 34, 56025 Pontedera, Italy; mauro.gemmi@iit.it
5 Department of Materials, University of Oxford, Oxford OX1 3PH, UK; angus.kirkland@materials.ox.ac.uk
* Correspondence: yu-jen.chou@mail.ntust.edu.tw (Y.-J.C.); konstantin.borisenko@kennedy.ox.ac.uk (K.B.B.)

Abstract: We investigated the influence of precession angle, energy filtering and sample thickness on the structural parameters of amorphous SiO_2 thin films from the electron reduced density functions obtained by applying precession electron diffraction. The results demonstrate that the peak positions in the electron reduced density functions are generally insensitive to the studied experimental conditions, while both precession angle and energy filtering influence peak heights considerably. It is also shown that introducing precession with small angles of up to 2 degrees and energy filtering results in higher coordination numbers that are closer to the expected theoretical values of 4 and 2 for Si and O, respectively, for data obtained from a thicker sample.

Keywords: transmission electron microscopy; precession electron diffraction; amorphous silica; reduced density function; pair distribution function

1. Introduction

Amorphous structures are an important class of materials for a number of applications, e.g., optical fibres [1] and biological scaffolds [2]. Understanding these materials' properties and building meaningful structure–property correlations depends on the accurate determination of atomistic structures. Due to the structural complexity of amorphous materials, describing an averaged building block or structure is one of the more generally used approaches to assess these structures [3]. Diffraction techniques are among the experimental techniques that are suitable for obtaining an averaged structure [4–6]. In this approach, the diffraction pattern can be converted into a pair distribution function (PDF) that describes an averaged structure, and this can be subsequently used to build an atomistic model.

Diffraction techniques have been applied extensively to determine atomic structures using various types of radiation including X-rays, neutrons and electrons. The major advantage of electron diffraction is that electrons have larger scattering cross sections (than X-rays), with the ability to form fine probes of controlled geometry. This enables the observation of the local atomic structure of materials from very small volumes when compared to those required for X-ray and neutron diffraction [7]. In addition, the imaging mode in a transmission electron microscope (TEM) allows for the straightforward selection of the sample area of interest from where the diffraction data are collected.

Electron reduced density function (eRDF) analysis, which is equivalent to pair distribution function (PDF) analysis, is a direct way to characterise the atomistic structure of amorphous thin films by recording and analysing diffraction patterns in a TEM.

The kinematic scattering approximation is often used when comparing atomistic models to experimental electron diffraction data, a practice which assumes that only one elastic scattering event occurs for each electron passing through the specimen [8]. However, in practice, the electrons can be scattered multiple times and scattering events can be inelastic, with characteristic energy losses.

The contribution of inelastic scattering can be removed by energy filtering [9]. Additionally, precession electron diffraction (PED) has been shown to reduce the effect of multiple scattering when interpreting diffraction data from crystalline samples [10–12].

The contributions of both multiple and inelastic scattering are expected to increase with sample thickness and therefore this is an important factor that will influence the results of an RDF analysis based on the kinematic scattering approximation.

Whilst eRDF analysis has been extensively applied in various studies [13–17], the effect of various experimental parameters on the eRDF analysis and especially on the refined atomistic models has not been systematically investigated for amorphous materials. The purpose of this study is to quantify the influence of precession angle in PED, together with the effects of energy filtering and sample thickness on the eRDF of amorphous SiO_2 thin films and, subsequently, on the determined structural parameters.

Electron Reduced Density Function (eRDF)

The total electron scattering intensity, $I(q)$, for a particular arrangement of different atomic species for all orientations of the sample can be calculated using Equation (1) [18], where $f(q)$ is the atomic scattering factor, q is the scattering vector and $r_{ij} = |r_i - r_j|$ is the distance between two atoms.

$$I(q) = \sum_i f_i^2(q) + \sum_{i \neq j} \sum_j f_i(q) f_j(q) \sin(q r_{ij}) / (q r_{ij}) \tag{1}$$

Following Cockayne [7], this function can be rewritten in terms of a pair distribution function, $g(r) = 4\pi r^2 \rho dr$, by integrating over the sample volume which is directly related to the local average density, (Equation (2)).

$$I(q) = \sum_i f_i^2(q) + 4\pi \left(\sum_i f_i(q) \right)^2 \int_0^\infty [g(r) - \rho_0] \frac{r}{q} \sin(qr) dr \tag{2}$$

In Equation (2), the atomic scattering intensity term, $\sum_i f_i^2(q)$, represents the atoms scattering independently and the second term describes the deviation from the mean intensity when the local density at r deviates from the mean density ρ_0. Used to extract the RDF, the reduced scattering intensity function $\varphi(q)$ is defined as in Equation (3).

$$\varphi(q) = \frac{[I(q) - \sum_i f_i^2(q)] q}{(\sum_i f_i(q))^2} \tag{3}$$

Finally, the function $\varphi(q)$ is Fourier-transformed to give the eRDF, (Equation (4)).

$$G(r) = 4\pi r [g(r) - \rho_0] = 4 \int_0^\infty \varphi(q) \sin(qr) dq \tag{4}$$

In practice, the eRDF (Equation (5)) is calculated by the integration of the reduced scattering function over an experimentally limited range of q between q_{min} and q_{max}. In this case, to reduce the influence of data truncation on the Fourier transform, the data are multiplied by an exponential function with a small artificial temperature factor b, which is usually set at $b = 0.02$:

$$G(r) = 4\int_{q_{min}}^{q_{max}} \varphi(q)\sin(qr)\exp\left(-bq^2\right)dq \quad (5)$$

2. Materials and Methods

2.1. Experimental

Electron diffraction data were collected from amorphous SiO_2 films samples with two different thickness of 234 Å and 471 Å (purchased from Agar Scientific Ltd., Essex, UK). The thicknesses were determined using an ARM-200CF TEM (JEOL, Akishima, Japan) from the field of electron energy loss spectroscopy (EELS) at 200 kV using the log–ratio technique [19] by averaging over three windows in each film sample. A Libra 120 TEM (Zeiss, Oberkochen, Germany), operating at 120 kV with an in-column energy filter, was used to collect diffraction data and electron diffraction patterns were recorded on a TRS full-frame 16-bit 2 k × 2 K-pixel CCD camera, with a camera length of 144 mm and a 5 μ condenser aperture giving a parallel nanobeam of 150 nm in diameter. The diffraction patterns were energy-filtered at the zero-loss peak, with a slit width of 20 eV. A NanoMEGAS Digistar P1000 was used for precession electron diffraction. When collecting diffraction patterns, a beam stop was used to block the central beam in order to collect scattering intensity data at larger scattering angles while avoiding the risk of damaging the CCD during long exposures.

2.2. eRDF Analysis

The calculation of eRDF curves from the experimental data was performed using the free open source eRDF Analyser software [20], which is one of several similar software suites available, including ProcessDiffraction [21], SUePDF [22], ePDF tools [23], and RDF Tools [24].

The eRDF Analyser provides an interactive graphical user interface and allows users to mask beam stops or any distorted areas, which are then excluded from the calculation of the final results. It also allows for the optimisation of the position of the centre of the diffraction pattern in the experimental diffraction data. Identifying the position of the centre is important for computing an accurate azimuthally averaged intensity distribution function, $I(q)$.

A polycrystalline aluminum specimen (SPI Supplies, West Chester, PA, USA) was used as a standard to calibrate the diffraction patterns. A customised automated fitting routine provided in the eRDF Analyser software was used to extract the atomic scattering background for calculations of the reduced scattering intensities according to Equation (3). This was performed based on least-squares fitting of the atomic scattering curve, with either the full range or the last third of the experimental data selected as the region of fitting. The reduced scattering intensity function $\varphi(q)$ was obtained from Equation (3), and the eRDF curve was obtained by a Fourier transform of $\varphi(q)$ following Equation (5). These functions were subsequently used in a reverse Monte Carlo refinement [25] to obtain an atomistic model, from which average coordination numbers and average bond lengths were calculated. The reduced scattering functions were obtained in the same range from 1.541 to 18.355 Å^{-1}, with a 0.014 Å^{-1} step to facilitate comparison between experiments.

2.3. Structure Refinement

Model structures were obtained from matched reduced scattering functions $\varphi(q)$ calculated from the model using Equation (6). Then, they were filtered to remove the contribution of frequencies, resulting in unphysically short distances in the RDF below 1 Å to the corresponding experimental scattering intensities, as described by McGreevy [25] and performed previously [26].

$$\varphi(q) = \frac{q \sum_{i \neq j} \sum_j f_i(q) f_j(q) \sin(qr_{ij}) / (qr_{ij})}{\left(\sum_i f_i(q)\right)^2} \quad (6)$$

Initially, a small model of SiO_2 with 102 atoms was prepared by random dense packing of atoms approximated as soft spheres, in a cubic cell with a lattice parameter corresponding to a density of 2.3 g cm^{-3}.

DFT calculations using the CASTEP code [27] were used to perform liquid quench molecular dynamics (MD) simulations using on-the-fly-generated pseudopotentials within the generalized gradient approximation using the Perdew–Burke–Ernzerhof exchange–correlation functional [28]. The generated pseudopotentials are a function of atomic species and, once generated in the beginning of the calculations, do not change during simulations. MD simulations were performed using a canonical (constant NVT) ensemble. The starting model was first melted at 3000 K for 2 ps. It was then simulated at 2000 K for 10 ps, and then further quenched to 400 K, reducing the temperature by 200 K in 5 ps equilibration steps. Finally, the structure was equilibrated at 300 K for 10 ps. The total simulation time was therefore 62 ps, with time steps of 2 fs throughout. The temperature was controlled using an implementation of the Nosé–Hoover thermostat [29]. The electron density was only sampled at the gamma point, with the pseudopotential energy cut-off set at 300 eV.

A larger model of 816 atoms for SiO_2 was constructed from the smaller model, obtained in the DFT MD simulations at 300 K, by assembling 2 × 2 × 2 small models with random orientations. These larger models were then used in RMC refinements. Small displacements of atoms in each of the RMC refinement steps were used to retain the structural features from the MD simulation, such as the preferred bond types and bond angle distributions. An example of a refined model is shown in Figure 1

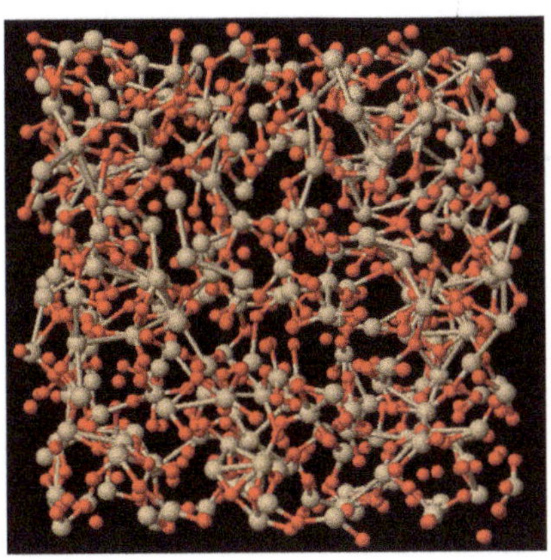

Figure 1. An example of SiO_2 model of 816 atoms after RMC refinement.

Unfiltered and filtered reduced scattering intensity functions are shown in Figure 2a. The filter is a low-order polynomial contribution to the background that cannot be fit to the refinements as it would mean the presence of unphysically close interatomic distances [25]. A comparison of the RDF obtained from the refined model to the experimental one computed from filtered scattering intensities is shown in Figure 2b.

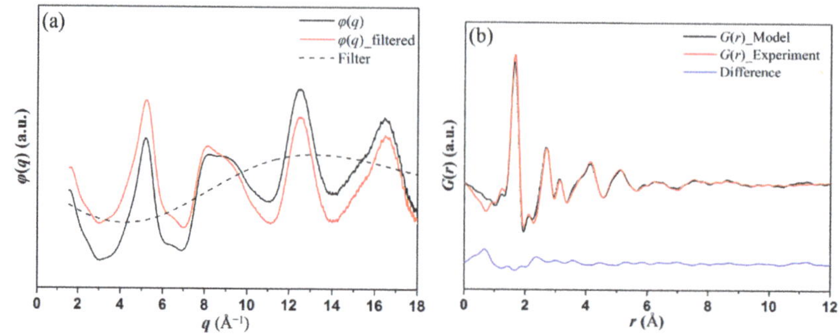

Figure 2. (**a**) $\varphi(q)$ filtering and (**b**) comparison between experimental and model $G(r)$ curves.

3. Results

Figure 3 shows a typical selected area diffraction pattern acquired from polycrystalline aluminium and amorphous SiO_2 films without energy filtering. The patterns were obtained with the central beam shifted to the corner of the detector to maximize the recorded scattering angles. The thinnest part of the beam stopper was used to block the unscattered beam in order to collect the scattered intensity at low angles. A pixel calibration factor of 0.00223 $Å^{-1}$ was determined using the eRDF Analyser by first identifying and refining the centre of the diffraction pattern obtained without energy filtering and then calculating $I(q)$ and relating the peak positions to the known reflections of polycrystalline Al, as summarised in Table 1.

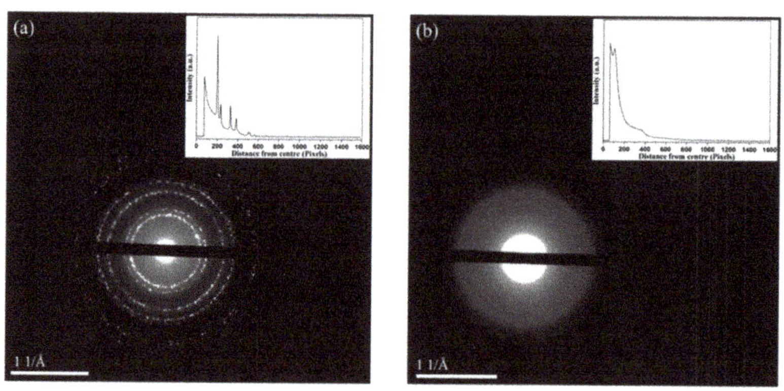

Figure 3. Typical electron diffraction patterns of (**a**) polycrystalline Al and (**b**) amorphous SiO_2 specimen with azimuthally averaged total scattering intensities shown inset.

Table 1. Computed pixel calibration factors for polycrystalline Al.

{hkl}	Pixel Number n (Pixel)	Lattice Spacing d (Å)	Calibration Factor (1/(*n*·*d*))
{111}	382	2.34	0.00224
{200}	445	2.02	0.00222
{220}	628	1.43	0.00223
{311}	734	1.22	0.00223
			Average = 0.00223

3.1. Influence of Specimen Thickness

Figure 4 compares the results of atomic scattering background fitting to the total experimental scattered intensity curves $I(q)$, $\varphi(q)$ and $G(r)$, recorded from amorphous

SiO_2 specimens with thicknesses of 234 Å and 471 Å, in experiments with 0° precession (no precession) and without energy filtering. Noticeable differences are observed in scattering intensities (Figure 4) at low scattering angles (q from 2 to 6 $Å^{-1}$) when comparing the data from the 471 Å film to the 234 Å data. This is due to the atomic scattering background for thinner specimens fitting better through the oscillations of the total scattered intensity curve in this region due to there being a smaller contribution from multiple and inelastic scattering, as seen in Figure 4. In the $G(r)$ curves, the first peak corresponds to the Si-O distances, while the second and third peaks are due to a mixture of Si ... O, Si ... Si and O ... O distances. The determined Si-O bond length of 1.61–1.63 Å also agrees with the value of 1.612(7) Å drawn from an X-ray diffraction study [30]. Measurements of peak position and height are given in Table 2. The maximum difference in peak positions for the two sample thicknesses is observed for the second peak and is about 0.03 Å, or less than 2%. The peak heights, however, show larger deviations for the two films, being of 12.6%, 6.4% and 21.9% for the first, second and third peaks, respectively. The peak heights were defined as the peak amplitude in the calculated eRDF curves. The significance of large change in the peak heights is that it affects the coordination numbers that can be calculated from the results.

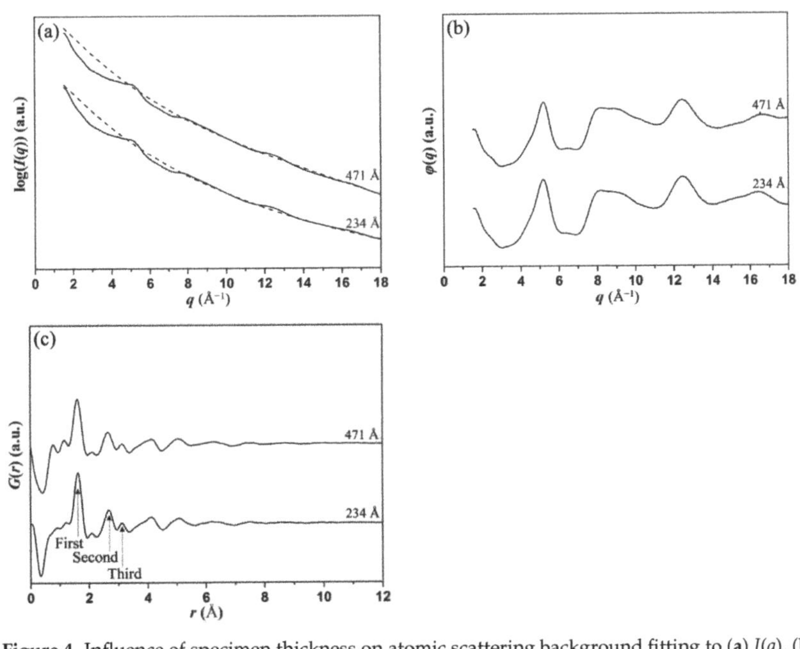

Figure 4. Influence of specimen thickness on atomic scattering background fitting to (**a**) $I(q)$, (**b**) $\varphi(q)$ and (**c**) $G(r)$ for samples with two different thicknesses, 234 Å and 471 Å.

Table 2. Influence of specimen thickness on (**a**) positions and (**b**) heights of the eRDF peaks.

(a) Peak Position/(Å)	First	Second	Third
234 Å	1.63	2.67	3.12
471 Å	1.61	2.64	3.13
(b) Peak Amplitude/(a.u.)	**First**	**Second**	**Third**
234 Å	2.70	1.10	0.32
471 Å	2.36	1.03	0.25

3.2. Influence of Precession Angle

PED was measured with precession angles from 0° to 3° for samples of both thicknesses and the resulting $\varphi(q)$ and $G(r)$ for both 234 Å and 471 Å samples are shown in Figure 5. The main parameters are given in Table 3. These results show that only small variations are observed when comparing the data without precession (0°) to those with precession (1° to 2°). However, when the precession angle reaches 3°, $\varphi(q)$ data at high q (above 16 Å^{-1}) appears larger, which leads to larger oscillations in the resulting $G(r)$. Furthermore, it can be seen from Table 3 that the peak positions differ very little when the precession angles vary from 0° (no precession) to 2°, with differences being less than 0.02 Å. However, for the 3° precession, the third peak position decreases by 0.08 Å. The same trend is observed for the peak heights. While experiments with precession angles from 0° to 2° show small differences of around 6.3% for all peaks, the 3° precession shows larger differences of 13.3%, 13.6%, and 65.6% for the first, second and third peaks, respectively. Similar trends are observed for the peak heights for the thicker sample of 471 Å with differences of 10.9%, 18.4%, and 47.1% for the first, second and third peaks, respectively, of the sample with 3° precession. This can be understood as the principle that applying large precession angle effectively increases the thickness of the sample. As suggested previously [31], the effect of sample thickness on the scattered intensity can be regarded as a convolution of the scattered intensity from a thin slice with itself as the electron wave propagates through a thicker sample. This results in a reduction of the amplitudes of all peaks in the total scattering intensity and, therefore, reduction in peak heights in G(r). Importantly, the large change in the peak heights has consequences for the determination of the coordination numbers from the results.

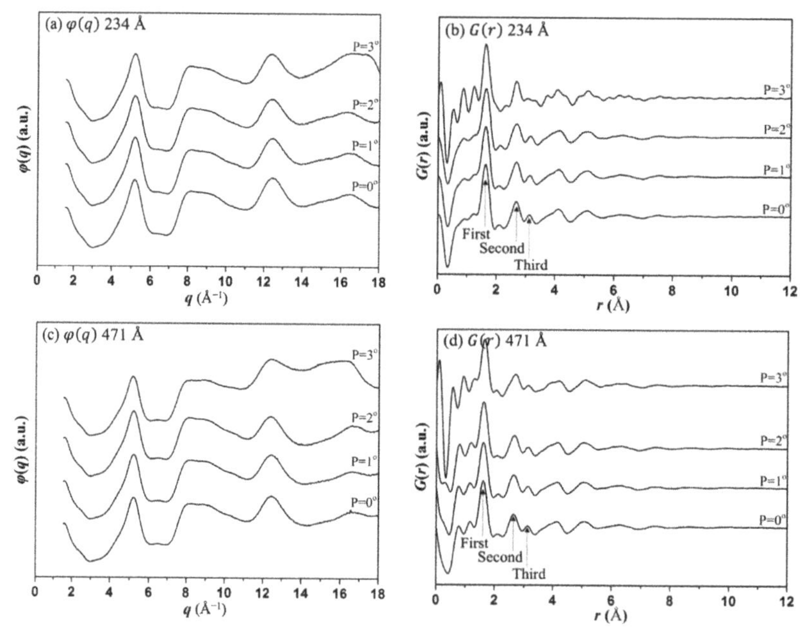

Figure 5. Influence of precession angle on $\varphi(q)$ and $G(r)$ for (a,b) 234 Å and (c,d) 471 Å samples.

Table 3. Influence of precession angle on (**a**) positions and (**b**) heights of the eRDF peaks for the 234 Å and 471 Å samples.

(a) Peak Position/Å		First	Second	Third
234 Å	P = 0°	1.63	2.67	3.12
	P = 1°	1.62	2.67	3.12
	P = 2°	1.63	2.67	3.14
	P = 3°	1.61	2.66	3.04
471 Å	P = 0°	1.61	2.64	3.13
	P = 1°	1.61	2.65	3.14
	P = 2°	1.62	2.66	3.13
	P = 3°	1.64	2.69	3.13
(b) Peak Amplitude/(a.u.)		**First**	**Second**	**Third**
234 Å	P = 0°	2.70	1.10	0.32
	P = 1°	2.66	1.11	0.34
	P = 2°	2.59	1.06	0.30
	P = 3°	3.06	1.25	0.11
471 Å	P = 0°	2.36	1.03	0.25
	P = 1°	2.31	1.03	0.27
	P = 2°	2.39	1.02	0.28
	P = 3°	2.65	0.87	0.17

3.3. *Influence of Energy Filtering on the RDF*

Figure 6 shows $\varphi(q)$ and $G(r)$ measured with and without energy filtering for both 234 Å and 471 Å samples. Although the overall shapes of $\varphi(q)$ are similar with and without energy filtering for both thicknesses, noticeable differences can be observed in curve oscillation amplitudes. The use of higher-amplitude oscillations in the $\varphi(q)$ in the experiments with energy filtering led to higher peaks in the corresponding $G(r)$ curves. Measurements of the eRDF curves parameters are shown in Table 4. It can be seen that peak positions in both $G(r)$ curves are identical, except for a small 0.01 Å difference for the first (Si-O) peak. However, the peak heights are 30.0%, 27.3%, and 56.3% smaller for the first, second and third peaks, respectively, when recorded without energy filtering. The trends for the 471 Å thick sample are similar. The difference in peak positions is small, being less than 0.02 Å, while the peak heights are 25.8%, 29.5%, and 13.8% smaller for the first, second and third peaks, respectively, without energy filtering. As the energy of the electrons defines their wavelength, the scattering of electrons of different energies will result in slightly shifted (relative to the incident beam) scattering profiles. Averaging over such scattering profiles at the detector is expected to have the effect of flattening the peaks, similar to the effects of the multiple scattering.

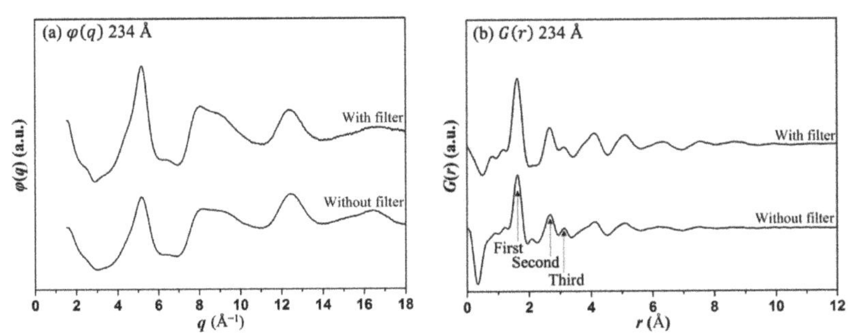

Figure 6. *Cont.*

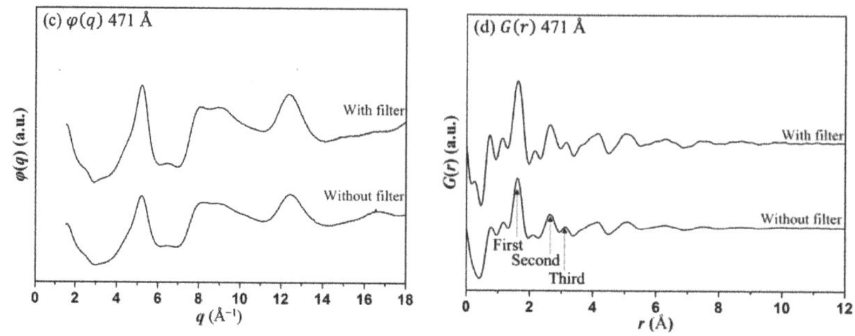

Figure 6. Influence of energy filtering on $\varphi(q)$ and $G(r)$ for (**a**,**b**) 234 Å and (**c**,**d**) 471 Å samples.

Table 4. Influence of energy filtering on (**a**) positions and (**b**) height of the eRDF peaks for 234 Å and 471 Å samples.

(a) Peak Position/(Å)		First	Second	Third
234 Å	Without energy filter	1.63	2.67	3.12
	With energy filter	1.62	2.67	3.12
471 Å	Without energy filter	1.61	2.64	3.13
	With energy filter	1.62	2.63	3.15
(b) Peak Amplitude/(a.u.)		**First**	**Second**	**Third**
234 Å	Without energy filter	2.70	1.10	0.32
	With energy filter	3.51	1.40	0.14
471 Å	Without energy filter	2.36	1.03	0.25
	With energy filter	3.18	1.46	0.29

3.4. Influence of Experimental Parameters on the Refined Structure

Table 5 summarises the coordination numbers, Z, and average Si-O bond length computed from the refined models of the amorphous SiO_2 films obtained under different experimental conditions, while Figure 7 shows the change in coordination numbers for various experimental conditions.

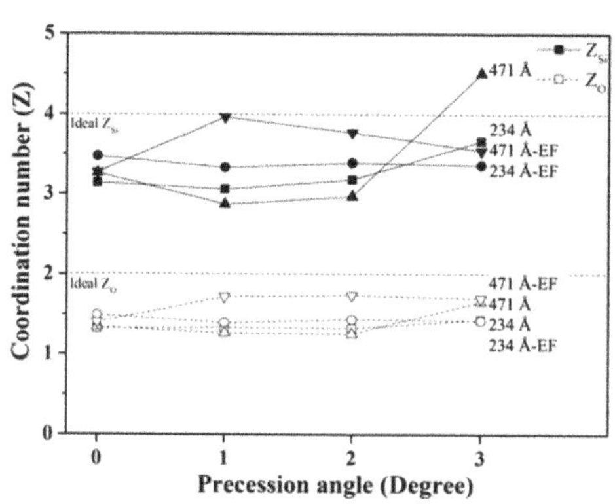

Figure 7. Variations in coordination numbers of Z_{Si} and Z_O with sample thickness, energy filtering and precession angle obtained from structure refinements of amorphous SiO_2 models.

Table 5. Influence of experimental parameters on refined coordination numbers and structural parameters of amorphous SiO_2 models.

Thickness/Å	Energy Filtering	Precession Angle/Deg	Z_{Si} (Ideal 4)	Z_O (Ideal 2)	Si-O/Å	R-Factor/%
234	No	0	3.14	1.32	1.67(19)	12.5
		1	3.06	1.33	1.68(20)	11.9
		2	3.18	1.32	1.67(19)	12.4
		3	3.66	1.43	1.63(19)	15.3
	Yes	0	3.47	1.49	1.64(18)	14.7
		1	3.33	1.39	1.64(15)	15.0
		2	3.39	1.43	1.64(14)	19.1
		3	3.36	1.42	1.64(12)	14.0
471	No	0	3.26	1.34	1.67(20)	12.8
		1	2.87	1.26	1.69(22)	12.5
		2	2.97	1.25	1.67(21)	15.7
		3	4.51	1.66	1.53(25)	19.7
	Yes	0	3.27	1.4	1.67(19)	16.8
		1	3.96	1.72	1.61(19)	16.9
		2	3.77	1.74	1.61(18)	17.6
		3	3.54	1.7	1.63(13)	16.1

4. Discussion

4.1. Influence of Sample Thickness

We have observed that the heights of all peaks are reduced with increasing thickness independently of precession or energy filtering, while the peak positions remain the same.

Increasing the sample thickness decreases the quality of fit, as seen in the R-factors in Table 5. This is to be expected as the contribution of inelastic and multiple scattering increases with sample thickness. This observation also agrees with earlier research showing that multiple scattering does not affect peak positions [31].

Although, the peak heights generally reduce with thickness, (Table 2) there is a less pronounced trend in the change in the coordination number obtained from the refined models (Figure 7). It can be seen, however, that the coordination number increases with thickness when energy filtering and precession are applied. Without energy filtering, the coordination number tends to be lower for the thicker sample when using smaller precession angles (Figure 7).

4.2. Influence of Precession

Smaller precession angles (1°, 2°) slightly decrease the height of the first peak as compared to data recorded without precession (Table 3), whilst a higher precession angle (3°) increases the height of the first peak considerably while notably reducing the height of the third peak. Its influence on fit quality is not straightforward. Introduction of small (1°) precession generally improves the fit quality. The improvement of fitting between model and experiment with the introduction of precession has also been seen in previous experiments with metallic nanoparticles [32]. The fit quality, however, reduces with larger precession angle. It can be suggested that larger precession angles effectively increase the sample thickness, which leads to larger contribution of inelastic scattering and, therefore, worse fit. This decrease in fit quality is notably larger for the thicker 471 Å sample (Table 5).

Other possible contributing factors cannot be excluded, such as an increased probability of descanning errors.

Using higher precession angles increased the coordination numbers for the thinner sample in the present investigation. For the thicker sample, however, a high degree of precession resulted in unreasonably high coordination number (4.5 for Si), which can be attributed to the worse fit due to effective increase in sample thickness and increased contribution of inelastic scattering at higher precession.

4.3. Influence of Energy Filtering

Energy filtering, as for precession, does not affect all peaks uniformly. For example, the heights of the first and second peaks increase considerably after energy filtering, (Table 4) as compared to the third peak, which reduces for the 234 Å thin sample or increases only slightly for the 471 Å thick one.

When energy filtering is introduced, coordination numbers increase for both sample thicknesses as compared to unfiltered data, but the goodness of fit reduces slightly (Table 5). The improvement in coordination number can be attributed to the removal of the inelastic scattering contribution by the energy filter and is a consequence of increased peak heights. The slight decrease in the fitting quality may be due to some experimental error introduced by the energy filter. The magnitude of change in coordination number with introduction of energy filtering is more notable for the thicker sample, as would be expected by the removal of the larger contribution of inelastic scattering.

5. Conclusions

The study demonstrates that the peak positions in the obtained eRDF curves are generally insensitive to sample thickness, precession angle or energy filtering.

However, the studied experimental conditions influence the peak heights considerably, which results in different coordination numbers being obtained from the atomistic models.

It is shown that introducing precession and energy filtering results in higher coordination numbers, bringing them closer to the ideal theoretical values for the material when the data are obtained from a thicker sample. It is also observed that energy filtering and precession, may influence the goodness-of-fit of atomistic models to the experimental data.

Author Contributions: Conceptualization: K.B.B., A.I.K. and S.N.; methodology: S.N. and M.G.; software: K.B.B.; investigation: Y.-J.C., M.G. and P.P.D.; resources: A.I.K. and S.N.; data curation: Y.-J.C. and K.B.B.; writing—original draft preparation: Y.-J.C.; writing—review and editing: Y.-J.C. and K.B.B.; supervision: A.I.K.; funding acquisition: A.I.K. All authors have read and agreed to the published version of the manuscript.

Funding: This research was funded by FP7-INFRASTRUCTURES grant number ESTEEM2 312483.

Data Availability Statement: The original experimental data are available from the authors upon request.

Acknowledgments: The authors are grateful to ESTEEM2 project for financial support.

Conflicts of Interest: The authors declare no conflict of interest.

References

1. Lee, B. Review of the present status of optical fiber sensors. *Opt. Fiber Technol.* **2003**, *9*, 57–79. [CrossRef]
2. Kuo, C.-K.; Huang, H.-W.; Chen, L.-G.; Chou, Y.-J. Fabrication and characterization of freeze dried strontium-doped bioactive glasses/chitosan composite scaffolds for biomedical engineering. *J. Asian Ceram. Soc.* **2021**, *9*, 1173–1182. [CrossRef]
3. Zachariasen, W.H. The atomic arrangement in glass. *J. Am. Chem. Soc.* **1932**, *54*, 3841–3851. [CrossRef]
4. Ferrari, A.C.; Robertson, J. Raman spectroscopy of amorphous, nanostructured, diamond–like carbon, and nanodiamond. *Philos. Trans. R. Soc. London. Ser. A Math. Phys. Eng. Sci.* **2004**, *362*, 2477–2512. [CrossRef]
5. Malfait, W.J.; Halter, W.E.; Morizet, Y.; Meier, B.H.; Verel, R. Structural control on bulk melt properties: Single and double quantum 29Si NMR spectroscopy on alkali-silicate glasses. *Geochim. Cosmochim. Acta* **2007**, *71*, 6002–6018. [CrossRef]
6. Cockayne, D.J.H.; McKenzie, D.R. Electron diffraction analysis of polycrystalline and amorphous thin films. *Acta Crystallogr. A* **1988**, *44*, 870–878. [CrossRef]

7. Cockayne, D.J.H. The Study of Nanovolumes of Amorphous Materials Using Electron Scattering. *Annu. Rev. Mater. Res.* **2007**, *37*, 159–187. [CrossRef]
8. Sproul, A.; McKenzie, D.R.; Cockayne, D.J.H. Structural study of hydrogenated amorphous silicon–carbon alloys. *Philos. Mag. B* **1986**, *54*, 113–131. [CrossRef]
9. Eggeman, A.S.; Barnard, J.S.; Midgley, P.A. Aberration-Corrected and Energy-Filtered Precession Electron Diffraction. *Z. Kristallogr.* **2013**, *228*, 43–50. [CrossRef]
10. Vincent, R.; Midgley, P.A. Double conical beam-rocking system for measurement of integrated electron diffraction intensities. *Ultramicroscopy* **1994**, *53*, 271–282. [CrossRef]
11. Gjønnes, K. On the integration of electron diffraction intensities in the Vincent-Midgley precession technique. *Ultramicroscopy* **1997**, *69*, 1–11. [CrossRef]
12. Gemmi, M.; Righi, L.; Calestani, G.; Migliori, A.; Speghini, A.; Santarosa, M.; Bettinelli, M. Structure determination of φ-Bi8Pb5O17 by electron and powder X-ray diffraction. *Ultramicroscopy* **2000**, *84*, 133–142. [CrossRef]
13. Abeykoon, A.M.; Hu, H.; Wu, L.; Zhu, Y.; Billinge, S.J. Calibration and data collection protocols for reliable lattice parameter values in electron pair distribution function studies. *J. Appl. Crystallogr.* **2015**, *48*, 244–251. [CrossRef]
14. Abeykoon, A.M.M.; Malliakas, C.D.; Juhás, P.; Bozin, E.S.; Kanatzidis, M.G.; Billinge, S.J.L. Quantitative Nanostructure Characterization Using Atomic Pair Distribution Functions Obtained From Laboratory Electron Microscopes. *Z. Kristallogr.* **2012**, *227*, 248–256. [CrossRef]
15. Kis, V.R.C.K.; Dódony, I.N.; Lábár, J.N.L. Amorphous and partly ordered structures in SiO_2 rich volcanic glasses. An ED study. *Eur. J. Mineral.* **2006**, *18*, 745–752. [CrossRef]
16. Petersen, T.C.; McBride, W.; McCulloch, D.G.; Snook, I.K.; Yarovsky, I. Refinements in the collection of energy filtered diffraction patterns from disordered materials. *Ultramicroscopy* **2005**, *103*, 275–283. [CrossRef]
17. Ankele, J.; Mayer, J.; Lamparter, P.; Steeb, S. Quantitative Electron Diffraction Data of Amorphous Materials. *Z. Für Nat. A* **2005**, *60*, 459–468. [CrossRef]
18. Warren, B.E. *X-ray Diffraction*; Courier Dover Publications: New York, NY, USA, 1969.
19. Malis, T.; Cheng, S.C.; Egerton, R.F. EELS log-ratio technique for specimen-thickness measurement in the TEM. *J. Electron Microsc. Tech.* **1988**, *8*, 193–200. [CrossRef]
20. Shanmugam, J.; Borisenko, K.B.; Chou, Y.-J.; Kirkland, A.I. eRDF Analyser: An interactive GUI for electron reduced density function analysis. *SoftwareX* **2017**, *6*, 185–192. [CrossRef]
21. Lábár, J.L.; Das, P.P. Pattern Center and Distortion Determined from Faint, Diffuse Electron Diffraction Rings from Amorphous Materials. *Microsc. Micoanal.* **2017**, *23*, 647–660. [CrossRef]
22. Tran, D.T.; Svensson, G.; Tai, C.-W. SUePDF: A program to obtain quantitative pair distribution functions from electron diffraction data. *J. Appl. Crystallogr.* **2017**, *50*, 304–312. [CrossRef]
23. Shi, H.; Luo, M.; Wang, W. ePDF tools, a processing and analysis package of the atomic pair distribution function for electron diffraction. *Comput. Phys. Commun.* **2019**, *238*, 295–301. [CrossRef]
24. Mitchell, D.R.G.; Petersen, T.C. RDFTools: A software tool for quantifying short-range ordering in amorphous materials. *Microsc. Res. Tech.* **2012**, *75*, 153–163. [CrossRef]
25. McGreevy, R.L. Reverse Monte Carlo modelling. *J. Phys. Condens. Mat.* **2001**, *13*, R877–R913. [CrossRef]
26. Borisenko, K.B.; Chen, Y.; Cockayne, D.J.H.; Song, S.A.; Jeong, H.S. Understanding atomic structures of amorphous C-doped Ge2Sb2Te5 phase-change memory materials. *Acta Mater.* **2011**, *59*, 4335–4342. [CrossRef]
27. Clark, S.J.; Segall, M.D.; Pickard, C.J.; Hasnip, P.J.; Probert, M.I.J.; Refson, K.; Payne, M.C. First principles methods using CASTEP. *Z. Für Krist.-Cryst. Mater.* **2005**, *220*, 567–570. [CrossRef]
28. Perdew, J.P.; Burke, K.; Ernzerhof, M. Generalized Gradient Approximation Made Simple. *Phys. Rev. Lett.* **1996**, *77*, 3865–3868. [CrossRef]
29. Martyna, G.J.; Klein, M.L.; Tuckerman, M. Nosé–Hoover chains: The canonical ensemble via continuous dynamics. *J. Chem. Phys.* **1992**, *97*, 2635–2643. [CrossRef]
30. Poulsen, H.F.; Neuefeind, J.; Neumann, H.B.; Schneider, J.R.; Zeidler, M.D. Amorphous silica studied by high energy X-ray diffraction. *J. Non-Cryst. Solids* **1995**, *188*, 63–74. [CrossRef]
31. Anstis, G.R.; Liu, Z.; Lake, M. Investigation of amorphous materials by electron diffraction—The effects of multiple scattering. *Ultramicroscopy* **1988**, *26*, 65–69. [CrossRef]
32. Hoque, M.M.; Vergara, S.; Das, P.P.; Ugarte, D.; Santiago, U.; Kumara, C.; Whetten, R.L.; Dass, A.; Ponce, A. Structural Analysis of Ligand-Protected Smaller Metallic Nanocrystals by Atomic Pair Distribution Function under Precession Electron Diffraction. *J. Phys. Chem. C* **2019**, *123*, 19894–19902. [CrossRef]

symmetry

MDPI

Article

Scanning Precession Electron Tomography (SPET) for Structural Analysis of Thin Films along Their Thickness

Sara Passuti, Julien Varignon, Adrian David and Philippe Boullay *

CRISMAT, CNRS , Normandie University, ENSICAEN, UNICAEN, 14000 Caen, France;
sara.passuti@ensicaen.fr (S.P.); julien.varignon@ensicaen.fr (J.V.); adrian.david@ensicaen.fr (A.D.)
* Correspondence: philippe.boullay@ensicaen.fr

Abstract: Accurate structure analysis of epitaxial perovskite thin films is a fundamental step towards the ability to tune their physical properties as desired. Precession-assisted electron diffraction tomography (PEDT) has proven to be an effective technique for performing *ab initio* structure solutions and refinements for this class of materials. As the film thickness or the region of interest (ROI) decrease in size, the capacity to collect PEDT data with smaller electron beams is a key parameter and ROI tracking becomes a major issue. To circumvent this problem, we considered here an alternative approach to acquiring data by combining PEDT with a scan over an area, extracting the intensities collected at different positions and using them to perform accurate structure refinements. As a proof of concept, a Scanning Precession Electron Tomography (SPET) experiment is performed on a 35 nm thick perovskite $PrVO_3$(PVO) film deposited on a $SrTiO_3$ (STO) substrate. This way, it was possible to detect small changes in the PVO structure along the film thickness, from the variation in unit cell parameters to atomic positions. We believe that SPET has the potential to become the standard procedure for the accurate structure analysis of ROIs as small as 10 nm.

Keywords: 3D ED; electron diffraction; thin films; perovskite

1. Introduction

Diffraction techniques face challenges in characterizing the crystal structures of nanomaterials. Epitaxial thin films of functional oxides serve as a prime example, with constant studies of ABO_3 perovskite-related compounds due to their various properties, such as ferroelectricity, magnetism and insulator–metal transitions [1–3]. Strain engineering can modify the properties of thin films by inducing small structural changes. The ensuing challenge is how to accurately describe small structural variations within thin films, typically in relation to bulk reference compounds. X-ray diffraction (XRD) is a widely used technique to measure epitaxial relationships, film metrics and assess whether the films are fully strained to the substrate or not. However, limitations arise from the standard acquisition configuration, where a significant portion of reciprocal space is inaccessible and substrate contributions are much larger than those of the film. Perovskite-based thin film structure refinement using XRD data is especially challenging for characterizing at a nanoscale the evolution of structural parameters such as the amplitude of BO_6 tilts and/or A displacements in the perovskite cage. These are parameters that may strongly affect the properties of perovskite compounds.

When considering structural crystallography methods, which involve refining a structural model using diffraction data, only electron diffraction (ED) provides the capability to probe structural alterations in regions of interest (ROIs) with dimensions significantly smaller than one micrometer. The interest in using ED, and more particularly 3D ED techniques [4], for the structure analysis of thin films has been largely demonstrated in some of our previous works where precession-assisted electron diffraction tomography (PEDT) has been used. PEDT first demonstrated its ability to solve the structure of unknown compounds deposited as thin films [5], something difficult if not impossible to obtain by X-ray diffraction. By implementing dynamic theory in the calculation of diffracted intensities and accounting for precession motion [6], it is possible to obtain reliable and accurate structural refinement on thin films [7,8]. In the study conducted by Steciuk et al. [8], precise refinements were accomplished on a 450 nm $CaTiO_3$ thick film using a beam size of approximately 60 nm, already with the idea to characterize structural evolution at different ROIs along the thickness. However, the contribution of several oriented domains (twins) in the diffracted intensities proved to be unavoidable and posed a significant challenge in obtaining accurate results if not accounted for.

For thin films well below 100 nm thickness, the ability to collect PEDT data with smaller electron beams is key to access relevant structural information. With advances in transmission electron microscopy (TEM) functionalities, PEDT analysis using a 10 nm (or smaller) electron beam is now possible. This is paving the way for accurate full-thickness film analysis. Tracking issues will then be a major issue when using such small beams. An efficient way to perform such an analysis is to scan the electron beam across the thickness of the film, i.e., in a direction perpendicular to its surface, at each tilt angle during a PEDT acquisition (Figure 1). Eggeman et al. [9] actually exploited a similar approach for analyzing the domain volume and orientation in crystalline Ni-based superalloys and used the acronym SPET for scanning precession electron tomography data acquisition. This concept was also recently used by Rauch et al. [10] in order to reconstruct in 3D the different domains composing a sample using parallel beam illumination. A related technique, namely 4D-STEM, consisting of acquiring 2D diffraction patterns on a 2D grid of positions, is already in use for performing orientation mapping with convergent beam illumination on multidomain samples composed of known phases [11]. In these above-mentioned works, SPET or 4D-STEM are used for microstructure analysis and, essentially, map known crystalline phases and orientations over an area (2D) or, less often, on a volume (3D).

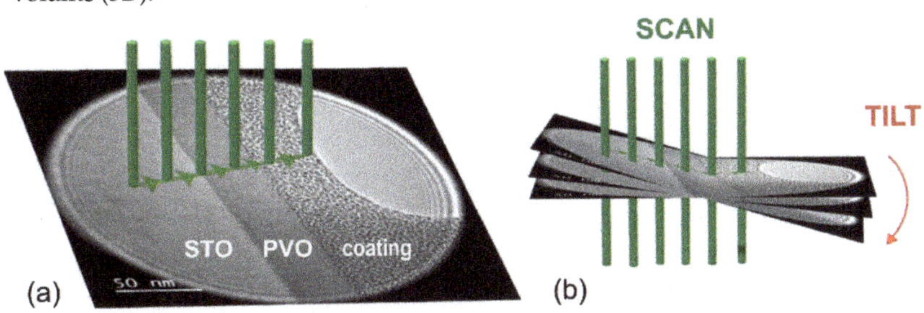

Figure 1. (**a**) Representation of the scanning procedure on a TEM image of the PVO thin film (dark-gray stripe). On its left and right sides we, respectively, find the STO substrate and the amorphous coating. The electron nanobeam while scanning along the thickness of the thin film is represented in green. After a scan is completed, the sample is tilted (**b**) and the scan is performed again.

In this study, we present preliminary results aimed at testing the potential of SPET for structure analysis and the outcomes of our investigation utilizing a 10 nm electron beam that traverses a 35 nm $PrVO_3$ (PVO) thin film deposited on a $SrTiO_3$ (STO) substrate. With a single SPET acquisition, we were able to analyze several ROIs throughout the thickness of the sample, achieve an ab initio structure solution and obtain accurate unit cell parameters and atomic coordinates for each one of the ROIs through dynamical refinements. As a result, we could observe small structural variations along the film thickness with enhanced spatial resolution compared to our prior works [8].

2. Materials and Methods

The object of our investigation is an epitaxial PVO[010]o thin film grown on STO[110]c substrate synthesized using pulsed-laser deposition as described in Kumar et al. [12], and analyzed in the form of TEM lamella. At room temperature, PVO has an orthorhombic crystal structure characterized by the space group *Pbnm* (SG: 62) and unit cell parameters a = 5.4856 Å, b = 5.5606 Å and c = 7.7771 Å. On the other hand, STO adopts a perovskite cubic structure with space group $Pm\bar{3}m$ (SG: 221), and its unit cell parameter is a = 3.905 Å. TEM imaging observations made on the same cross-sectional lamella utilized in this study and our prior work [12] confirmed that the film is similar to a single domain throughout its length and thickness. This is rather uncommon for epitaxial oxide thin films, where domains with different orientations are frequently present. However, we considered that this characteristic could serve as a model thin film system for testing the ability of SPET to access structural information over a sample area and, in the present case, the thickness of the film. Given that the bulk STO exhibits an $a^0b^0c^0$ tilting system in accordance with Glazer's notation [13], while bulk PVO has an $a^-b^-c^+$ tilt pattern, we anticipated observing a completely strained segment of the film close to the substrate, as depicted in Figure 2a. This implies, for instance, that the V-O-V angles of PVO would be close to 180°, as is the case in the perovskite structure of STO (Figure 2b). As we moved away from the substrate, we expected a relaxation of the strain imposed on PVO. Eventually, it returned to a configuration of the atoms close to the one observed in bulk PVO, including octahedral tilt amplitudes (Figure 2c).

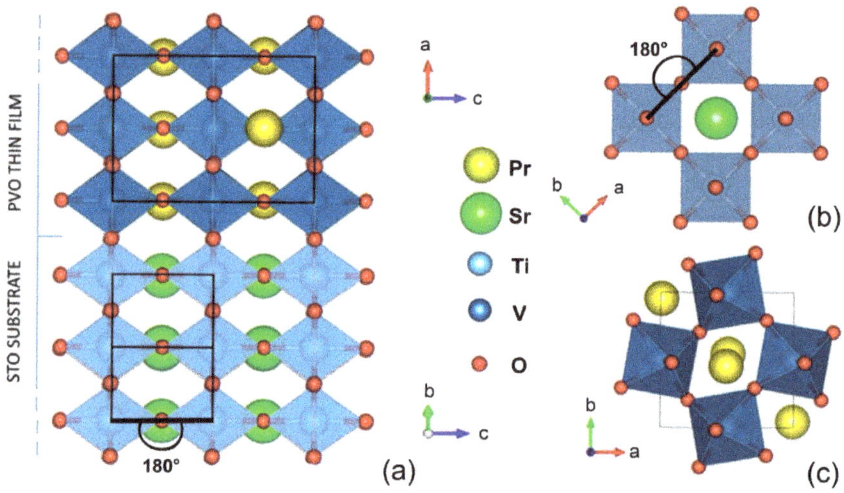

Figure 2. (**a**) Schematic representation of the expected strained structure of the PVO thin film on the STO substrate in the first deposited layers, viewed along the direction of the interface corresponding to the a axis of PVO. The unit cells of PVO and STO are represented in black. (**b**,**c**) Projections along the c axis of the STO and PVO unit cells in bulk, respectively, highlighting the difference in the tilting system of the BO_6 octahedra between the two.

SPET produces a series of PEDT data recorded at once on different ROIs. They were acquired using a JEOL F200 (200 kV) TEM equipped with a NanoMEGAS Digistar unit capable of generating simultaneous precession and line scan motions. A 10 nm diameter electron beam was scanned in a direction perpendicular to the interfaces, from the substrate to the coating with a step of approximately 0.92 nm between two acquisitions, resulting in oversampling of the same area of the film. The diffraction patterns were acquired with an ASI Cheetah M3 hybrid pixel detector (512 × 512 pixels) with 0.02 s exposure time. The precession semi-angle was set to 1.4°. A goniometer tilt step of 2° was applied after each line scan in the range [+44°,−34°], while a tilt step of 1° was used in the range [−34°,−50°]. To avoid shadowing of the film by the substrate or coating during sample rotation [8], the tilt axis was aligned with the direction of scanning. Importantly, no tracking of a specific area was performed between each goniometer tilt. A rapid visual examination was carried out to ensure that, for each tilt angle, the SPET data were systematically collected from the substrate to the coating.

For analysis, PEDT data from one given ROI were extracted from the SPET series and processed using the program PETS 2.0 [14]. In this work, only 9 ROIs (out of 38 collected) were selected for data processing as indicated in Table 1 and in Figure 3. As structural parameters change rapidly close to the substrate, more ROIs were taken into account in this part of the film. Each ROI is referred to in relation to the distance (in nm) from the substrate interface as presented Figure 3. Collecting PEDT data from both the substrate and coating during SPET data collection serves two purposes. Firstly, it indicates when the film is entered and exited during the scan. Secondly, PEDT data obtained from the substrate can act as an internal standard, enabling an accurate determination of the unit cell parameters. This is achievable thanks to new functionalities recently implemented in PETS 2.0 to correct distortions induced by the microscope's optical system and by the precession motion [15]. To ensure a standardized procedure for data analysis, we first determined calibration constant and distortion parameters on STO and used them for the subsequent analysis of PEDT data collected on different ROIs of the film through its thickness. Kinematical and dynamical refinements were performed on JANA2020 with the results of the dynamical refinements reported in Table 1. Note that while the kinematically derived R(obs) values typically fell within the range of 25 to 30%, the use of dynamical refinements resulted in a significant reduction in these values (Table 1), accompanied by an enhancement in the accuracy of structural parameters.

Table 1. SPET specifics and results for dynamical refinements of PVO at different thicknesses. Fixed coordinates correspond to atoms positioned at special positions of the Pbnm space group.

SPET Data Acquisition		
ROIs per scan in total/PVO 120/38	Tilt range 95° for 57 frames	PED semi-angle 1.4°
PEDT data analysis		
g_{max} = 1.6 Å^{-1}	RSg_{max} = 0.4	integration steps = 126
GoF(all) $^{min.}/_{max.}$ = $^{2.96}/_{3.84}$	Robs $^{min.}/_{max.}$ = $^{0.1247}/_{0.1610}$	wRall $^{min.}/_{max.}$ = $^{0.279}/_{0.3412}$

PVO atomic positions							
ROI (nm)	x(Pr1)	y(Pr1)	x(O1)	y(O1)	x(O2)	y(O2)	z(O2)
0.52	0.998(2)	0.0078(6)	0.019(4)	0.498(2)	0.741(2)	0.258(2)	0.006(6)
3.28	0.996(2)	0.0140(6)	0.019(6)	0.495(3)	0.736(3)	0.261(3)	0.014(4)
4.20	0.995(2)	0.0161(6)	0.020(6)	0.495(3)	0.735(3)	0.263(3)	0.016(4)
5.12	0.995(4)	0.0182(9)	0.022(5)	0.493(3)	0.734(3)	0.264(2)	0.018(4)
7.88	0.994(2)	0.0257(9)	0.040(6)	0.488(5)	0.727(3)	0.270(3)	0.025(5)
9.72	0.993(2)	0.031(1)	0.046(9)	0.492(5)	0.725(4)	0.276(3)	0.030(6)
12.48	0.993(1)	0.0375(9)	0.063(6)	0.485(5)	0.719(3)	0.281(3)	0.037(5)
17.08	0.9919(9)	0.0428(9)	0.070(5)	0.480(4)	0.715(3)	0.285(2)	0.036(4)
29.08	0.994(2)	0.038(1)	0.073(5)	0.491(4)	0.714(4)	0.282(3)	0.032(6)
All	*fixed:*	*z(Pr1) = 0.25*	*x(V1) = 0.5*	*y(V1) = 0.0*	*z(V1) = 0.0*	*z(O1) = 0.25*	

We performed *first-principles* simulations to corroborate the experimental lattice parameters and tilt angles obtained by SPET. For this, we used Density Functional Theory simulations using the Vienna Ab initio Simulation Package [16] in combination with the meta-Generalized Gradient Approximation (meta-GGA) Strongly Constrained and Appropriately Normalized (SCAN) exchange–correlation functional [17]. This functional was previously shown to produce the correct description in terms of the insulating or metallic character and structural parameters of various 3*d* transition metal oxide perovskites [18]. Projector-Augmented Wave (PAW) [19] potentials are used to model the core electrons and only the valence electrons are treated explicitly in the simulations—Pr 4*f* electrons are included in the potential. The initial structure supplied to the solver is a *Pbnm* cell, corresponding to a 4 f.u supercell with respect to the primitive high symmetry $Pm\bar{3}m$ cubic cell and characterized by an $a^-a^-c^+$ octahedral rotation pattern in Glazer's notation [20]. It is accompanied by an AFMC order consisting of AFM couplings between nearest V spins in the (ab)-plane and a FM coupling between consecutive planes along the c axis. This magnetic order is observed experimentally at low temperatures [21]. The structural relaxation is performed until forces acting on each atom are lower than 0.005 eV/Å. The epitaxial strain experienced by the film is finally modeled by imposing two lattice parameters of the PVO *Pbnm* cell to that of a (110)-oriented STO substrate and the third one is free to relax in amplitude and direction.

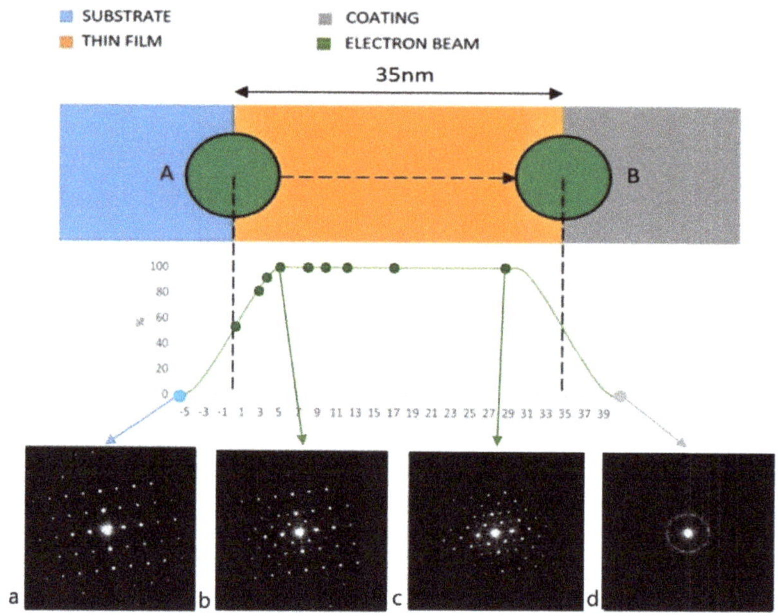

Figure 3. Schematic representation of the SPET experiment. The horizontal dotted arrow indicates the scanning direction, while A and B configurations of the electron beam coincide, respectively, with 0 and 35 in the nm scale. Beneath the scheme, a graph representing the percentage of electron beam area impinging the film for every position is shown, considering a beam diameter of 10 nm. Diffraction patterns extracted from the SPET experiment processed in this work at their respective thicknesses are indicated as green dots. Examples of diffraction patterns are also provided for (**a**) the substrate, (**b**) the thin film at 5 nm, (**c**) the film at 29 nm and (**d**) the coating.

3. Results

The first outcome is the ability to ascertain the epitaxial relationship by examining the 3D reconstructed reciprocal lattice from both the STO substrate and the PVO film. Using this direct output from 3D ED analysis, we confirm that the PVO film grows in the b direction (*Pbnm* orthorhombic cell), with the a and c directions lying on the interface with the STO substrate. As a result, we anticipate that the lattice parameters in the a and c directions of PVO will be significantly constrained, while the lattice parameter in the b direction may exhibit slight variations with respect to film thickness. However, determining the evolution of the lattice parameters as a function of film thickness using PEDT data remains a challenge. In our previous studies on thin films, this aspect was unaddressed due to the inherent limitations in determining the accuracy of lattice parameters. Thanks to the extensive efforts by P. Brazda et al. [15], this limitation has recently been overcome, enabling us to apply this to thin films, which serves as a valuable test case.

In Figure 4, the unit cell parameters of the PVO film, obtained from the analysis of SPET data, are presented for various thicknesses of the sample. As mentioned in the section Materials and Methods, STO was used as an internal standard for determining the calibration and distortion parameters. Doing so, and without any other constraint imposed during the data processing, we first notice that the *a* and *c* parameters of the PVO film exhibit no significant variation along the film thickness, in agreement with the fact that they are strained to match the unit cell of the substrate throughout the thickness, as supported by the Reciprocal Space Mapping (RSM) published in Kumar et al. [12]. At the interface with STO, there is no VO_6 octahedra rotation in PVO and the a and b parameters are expected to be equal, as observed. When we move away from the substrate, the structure of the film tends to return to a state close to the one adopted in the bulk material. The *b* parameter is the only one not directly strained by the substrate and shows a notable increase, well above the standard uncertainty range, reaching a plateau after 15 nm. This indicates that the out-of-plane parameter relaxes with increasing thickness in order to accommodate some structural changes within the PVO film. Globally, the obtained lattice parameters at the free surface of the film (Figure 4) do not match the ones observed for bulk PVO due to (i) the strain imposed by STO for a and c lattice parameters across the whole film and (ii) the progressive relaxation of octahedral rotations across the PVO film that produces an increase in the *b* lattice parameter, as we will show later.

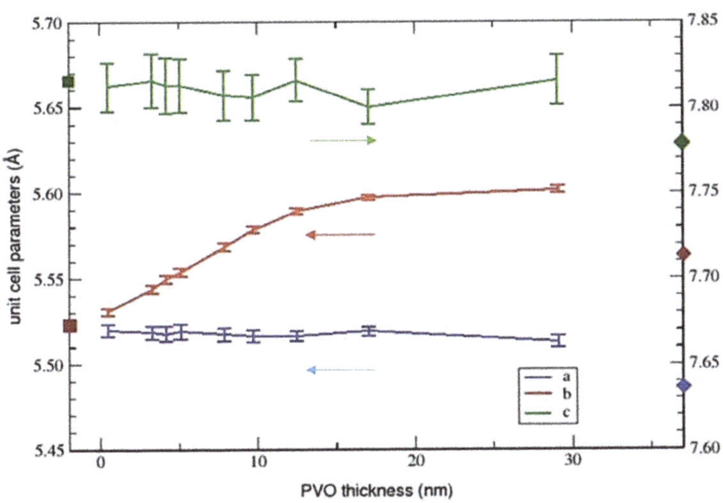

Figure 4. Refined unit cell parameters of PVO for different thicknesses of the thin film. The arrow under each experimental curve indicates the y scale to which the curve refers. The squares on the left y axis represent the reference unit cell parameter of bulk STO transformed in order to fit the orthorhombic cell of PVO ($a_{ortho} = b_{ortho} = a_0\sqrt{2}$; $c_{ortho} = 2 * c_0$). Since a_{ortho} and b_{ortho} are equivalent and superimposed, only one of the two is visible. The rhombus values on the right y axis represent the reference unit cell parameters of bulk PVO. Squares and rhombuses are of the color of the series they refer to.

While not being the primary objective of this study, it is noteworthy to mention that the PVO structure can be determined *ab initio* from the experimental intensities recorded using SPET. To know if the evolution of the lattice parameters as a function of the film thickness can actually be correlated to an evolution of the PVO structure, we performed dynamical refinements for different ROIs along the film thickness. Our focus was on two key structural features: the positioning of praseodymium within the perovskite cage and the rotation of the VO_6 octahedra. Regarding the position of Pr in the unit cell, as represented in Figure 5, only the x and y coordinates are considered since z is fixed by symmetry in the space group *Pbnm*. At the interface with STO, Pr is assumed to be in the center of the perovskite cage like Sr in STO with, here, coordinates x = 1 and y = 0 (square marks on the left *y* axis Figure 5). Looking at the Pr coordinates in bulk PVO (rhombuses marked on the right *y* axis Figure 5), the main difference is the y coordinate. Accordingly, while the x coordinate shows little variation across the film thickness, the y coordinate increases significantly to reach the bulk value. This variation is consistent with the strain relaxation along the film thickness observed for the *b* parameter. Similarly, the evolution is strong between 0 and 15 nm and small after 15 nm. The evolution of the V-O-V angles, as plotted in Figure 6, allows us to estimate the amplitude of the octahedral tilting across the film thickness. Close to STO, the substrate imposes a force onto the PVO structure so that V-O-V angles tend to be close to 180° (no rotation). Going away from the substrate, both the V-O1-V and V-O2-V angles show a decrease, i.e., an increment in the octahedral tilting amplitudes, up to 15 nm where a plateau is reached at about 156°, close to the values found in the bulk PVO structure.

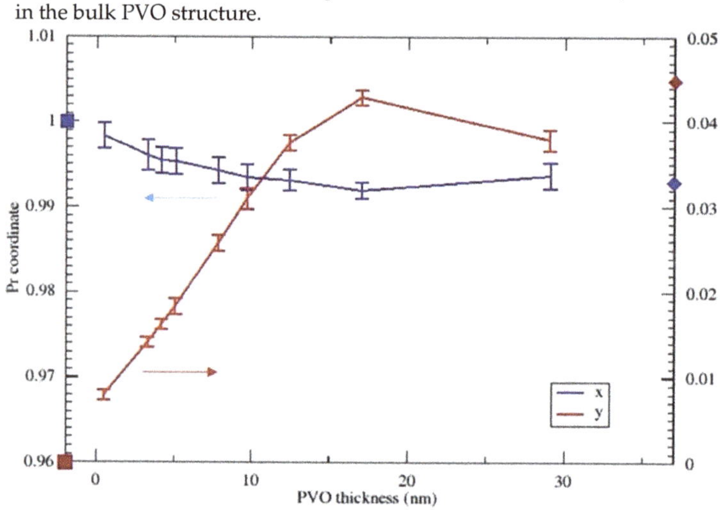

Figure 5. Refined coordinates of Pr atom in PVO unit cell for a selection of ROIs across the film thickness. The arrow under each experimental curve indicates the y scale to which the curve refers, and z values were not plotted since the Sr atom lies on the special position z = 0.25. Squares on the left y axis indicate the Sr coordinates for STO, corresponding in PVO to the atomic site of Pr, while rhombuses on the right y axis indicate the Pr coordinates for bulk PVO, the color corresponding to the one of the series they refer to.

The evolution of the lattice parameters and structural features obtained using SPET are, globally, consistent with what we could expect considering the strain relaxation where the film tries to recover a structure close to bulk PVO. If we look critically at these results, there are a few points that deserve attention. The first one is for the results obtained close to the interface with STO. As depicted in Figure 3, using a 10 nm beam, no information can be obtained directly for the very first deposited layers of PVO as the diffraction signal is largely dominated by the substrate contribution. Only an interpolation can be made between a nontilted perovskite at thickness 0 and the PEDT data we obtained for thicknesses above 5 nm, for which the beam is entirely in the film. The second point is how to interpret the results after 5 nm, where we can state that the obtained structural difference with respect to the PVO bulk structure is not to be traced back to a contribution of the STO substrate in the diffraction patterns.

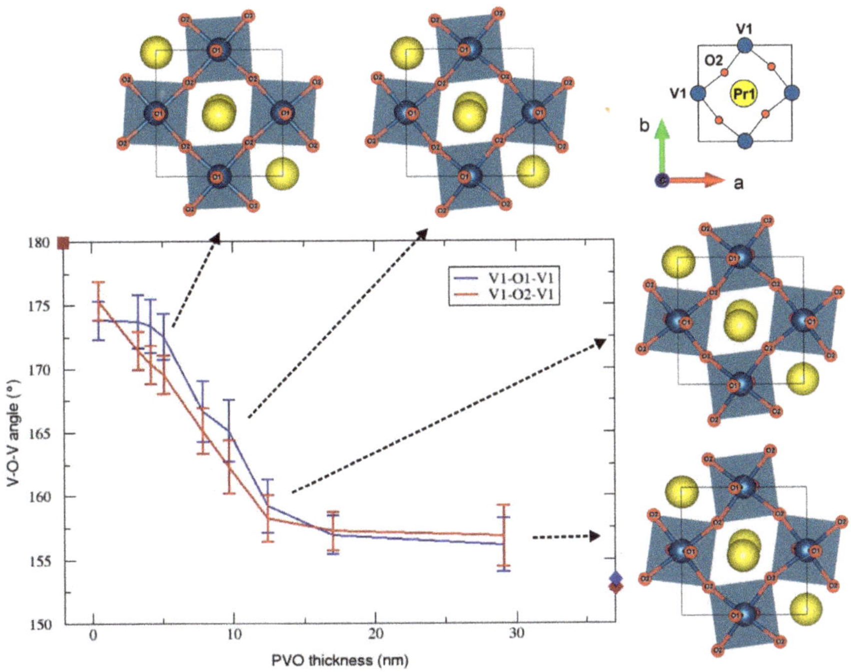

Figure 6. Angles for the series V1-O1-V1 and V1-O2-V1 obtained from dynamical refinements of PVO for different thicknesses of the thin film. Squares on the left y axis indicate the Ti-O-Ti angle for STO. Rhombuses on the right y axis indicate the V-O-V angles in bulk PVO, the color corresponding to the color of the series they refer to. To illustrate the changes in PVO, a projection of the structure along the c axis is given, specifically emphasizing the evolution of the V1-O2-V1 angles (as well as the Pr shift—see Figure 5).

In order to obtain insight into the reliability of the SPET method for extracting structural parameters across the thickness of thin films, DFT simulations were performed on a [010]-oriented PVO film with respect to the orthorhombic cell. After the structural relaxation performed at 0 K, we observe that two lattice parameters are elongated with respect to the 0K DFT bulk structure (i.e., $a_{strained}$ = 5.5225 Å and $c_{strained}$ = 7.8100 Å versus a_{bulk} = 5.4856 Å and c_{bulk} = 7.7771 Å), in agreement with the experimental measurement presented in Figure 4. This leads to a small contraction of the b lattice parameter in order to minimize the relative volume variation with respect to the bulk ($b_{strained}$ = 5.5230 Å versus b_{bulk} = 5.5606 Å). At first glance, this result disagrees with the experimental value at 293 K reported in Figure 4 (i.e., b = 5.56 Å). Nevertheless, this discrepancy may originate from the fact that our DFT simulations are performed at 0 K and a fair agreement is indeed recovered once using the experimental b parameter obtained at 5 K (b = 5.6061 Å [22]). Regarding the octahedral rotation amplitude, we extracted V-O1-V and V-O2-V angles of 152.3° and 154.1°, respectively. This indicates a slight reduction in octahedral rotation with respect to the DFT relaxed bulk in which the two angles are 151.4° and 153.0°. This trend agrees with the experimental observation for these two angles far away from the interface, where no substrate effect may be experienced by PVO (Figure 6).

One can actually relate the evolution of the b lattice parameter to the V-O-V angles (Figure 7). Remarkably, we notice an elongation of the b lattice parameter upon increasing the octahedral rotation amplitude. This behavior can be tracked theoretically by (i) fixing a fractional amplitude of the relaxed atomic positions with respect to the high symmetry cell from 0 (no distortion) to 1 (total distortion appearing in the strained material) and

(ii) performing relaxation of the *b* lattice parameter. The evolution of b/b_{max} with the octahedral rotation amplitude from DFT is reported in Figure 7. We observe that without any octahedral rotation, the *b* lattice parameter is expected to be rather small. From 0.25 to 0.75, a progressive elongation of *b* is observed. Above 0.75, *b* reaches a plateau with its maximal elongation. The overall trend of *b* versus the octahedral rotation amplitude is well reproduced by DFT, confirming the reliability of the experimental measurements.

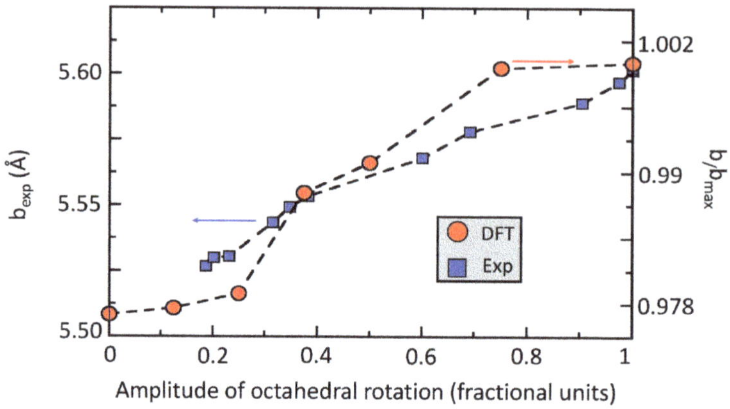

Figure 7. Evolution of the *b* lattice parameter (in Å) as a function of the octahedral rotation amplitude (in fractional units) from experiments (filled blue squares, left scale) and variations in b/b_{max} from DFT simulations (red filled circles, right panel). An amplitude of 1 corresponds to the largest value of octahedral rotation amplitude from theory and experiments, respectively. b_{max} is set to the largest value of *b* obtained for a rotation amplitude of 1 in the DFT simulations.

4. Conclusions

In this work, we introduced a new approach to thin film characterization based on 3D ED. SPET has been previously used [9,10] to obtain crystallographic phase and orientation mapping on a volume. In this study, we extracted electron diffraction intensities (3D ED datasets) from different ROIs and analyzed them quantitatively, aiming for accurate structure refinements. Specifically, SPET was used on a 35 nm PVO thin film, combining a scanning procedure with PEDT. By scanning with a nanobeam of 10 nm diameter along the PVO film thickness at every tilt step during a PEDT experiment, we could analyze the film structure at different thicknesses by properly sorting the acquired diffraction patterns.

Cell parameters and atomic positions were accurately determined and an evolution along the thickness of the film was observed in the out-of-plane unit cell parameter, as well as in the tilting of the VO_6 octahedra and in the y coordinate of the Pr atoms. These variations are coherent with the expected constraint given by the substrate in the first layers, and afterwards with the expected gradual relaxation of the crystal structure. At the interface with STO, oxygen atoms in PVO are pinned by the substrate and the angles are close to 180°, causing the smallest value of the *b* parameter. Progressively moving farther away from the interface, V-O-V angles go towards smaller values and then *b* increases. Finally, far away from the interface, the V-O-V angles exhibit their smallest values and hence *b* reaches its largest amplitude and a plateau. By comparing the experimental results and DFT calculations, we demonstrated that SPET in nanobeam configuration is a valid technique to obtain accurate information about subtle changes in the crystal structure of thin films. These results obtained by a 3D ED method at such a small scale are pioneering.

Since these variations are known to occur in the first few nm of thin films, also depending on the degree of constraint given by the substrate, it is fundamental to use, for such experiments, an electron beam diameter as small as possible in order for the results of the refinement to be as accurate as possible with respect to each probed region of the sample. Implementing the data acquisition strategy and improving the diffraction pattern sorting procedure, this analysis has the potential to become faster and even more accurate, paving the way for a new routine characterization for thin films. This can in turn lead to an important step forward in the field of thin film engineering, finally making it possible to link their physical properties to their accurately determined crystal structure.

Author Contributions: Conceptualization, P.B.; methodology, S.P. and P.B.; validation, all; formal analysis, S.P., J.V. and P.B.; investigation, S.P. and J.V.; resources, A.D.; data curation, S.P.; writing—original draft preparation, S.P. and P.B.; writing—review and editing, all; visualization, all; supervision, A.D. and P.B.; project administration, P.B.; funding acquisition, P.B. All authors have read and agreed to the published version of the manuscript.

Funding: This research was funded by the European Union's Horizon 2020 research and innovation program under the Marie Skłodowska-Curie grant agreement No. 956099 (NanED–Electron Nanocrystallography–H2020-MSCA-ITN).

Data Availability Statement: The 3D ED raw data will be deposited and findable in Zenodo NanED community (https://zenodo.org/communities/naned) (accessed on 12 May 2023).

Acknowledgments: J.V. acknowledges access granted to the HPC resources of Criann through the projects 2020005 and 2007013 and of Cines through the DARI project A0080911453.

Conflicts of Interest: The authors declare no conflict of interest.The funders had no role in the design of the study; in the collection, analyses, or interpretation of data; in the writing of the manuscript; or in the decision to publish the results.

References

1. Fontcuberta, J. Multiferroic $RMnO_3$ thin films. *CR Phys.* **2015**, *16*, 204–226. [CrossRef]
2. Bea, H.; Bibes, M.; Fusil, S.; Bouzehouane, K.; Jacquet, E.; Rode, K.; Bencok, P.; Barthélémy, A. Investigation on the origin of the magnetic moment of $BiFeO_3$ thin films by advanced X-ray characterizations. *Phys. Rev. B* **2006**, *74*, 020101. [CrossRef]
3. Ahn, K.H.; Lookman, T.; Bishop, A. Strain-induced metal-insulator phase coexistence in perovskite manganites. *Nature* **2004**, *428*, 401–404. [CrossRef]
4. Gemmi, M.; Mugnaioli, E.; Gorelik, T.E.; Kolb, U.; Palatinus, L.; Boullay, P.; Hovmöller, S.; Abrahams, J.P. 3D electron diffraction: The nanocrystallography revolution. *ACS Cent. Sci.* **2019**, *5*, 1315–1329. [CrossRef] [PubMed]
5. Zhang, W.; Li, M.; Chen, A.; Li, L.; Zhu, Y.; Xia, Z.; Lu, P.; Boullay, P.; Wu, L.; Zhu, Y.; et al. Two-Dimensional Layered Oxide Structures Tailored by Self Assembled Layer Stacking via Interfacial Strain. *ACS Appl. Mater. Interfaces* **2016**, *8*, 16845–16851. [CrossRef] [PubMed]
6. Palatinus, L.; Petříček, V.; Corrêa, C.A. Structure refinement using precession electron diffraction tomography and dynamical diffraction: Theory and implementation. *Acta Crystallogr. A* **2015**, *71*, 235–244. [CrossRef]
7. Veis, M.; Minár, J.; Steciuk, G.; Palatinus, L.; Rinaldi, C.; Cantoni, M.; Kriegner, D.; Tikuišis, K.K.; Hamrle, J.; Zahradník, M.; et al. Band structure of CuMnAs probed by optical and photoemission spectroscopy. *Phys. Rev. B* **2018**, *97*, 125109. [CrossRef]
8. Steciuk, G.; David, A.; Petříček, V.; Palatinus, L.; Mercey, B.; Prellier, W.; Pautrat, A.; Boullay, P. Precession electron diffraction tomography on twinned crystals: Application to $CaTiO_3$ thin films. *J. Appl. Crystallogr.* **2019**, *52*, 626–636. [CrossRef]
9. Eggeman, A.S.; Krakow, R.; Midgley, P.A. Scanning precession electron tomography for three-dimensional nanoscale orientation imaging and crystallographic analysis. *Nat. Commun.* **2015**, *6*, 7267. [CrossRef]
10. Rauch, E.F.; Harrison, P.; Zhou, X.; Herbig, M.; Ludwig, W.; Véron, M. New Features in Crystal Orientation and Phase Mapping for Transmission Electron Microscopy. *Symmetry* **2021**, *13*, 1675. [CrossRef]
11. Ophus, C. Four-dimensional scanning transmission electron microscopy (4D-STEM): From scanning nanodiffraction to ptychography and beyond. *Microsc. Microanal.* **2019**, *25*, 563–582. [CrossRef]
12. Kumar, D.; David, A.; Fouchet, A.; Pautrat, A.; Boullay, P.; Jung, C.U.; Prellier, W. Strong magnetic anisotropy of epitaxial $PrVO_3$ thin films on $SrTiO_3$ substrates with different orientations. *ACS Appl. Mater. Interfaces* **2020**, *12*, 35606–35613. [CrossRef] [PubMed]
13. Glazer, A.M. The Classification of Tilted Octahedra in Perovskites. *Acta Crystallogr. B* **1972**, *28*, 3384–3392. [CrossRef]
14. Palatinus, L.; Brázda, P.; Jelínek, M.; Hrdá, J.; Steciuk, G.; Klementová, M. Specifics of the data processing of precession electron diffraction tomography data and their implementation in the program PETS2. 0. *Acta Crystallogr. Sect. Struct. Sci. Cryst. Eng. Mater.* **2019**, *75*, 512–522. [CrossRef]
15. Brázda, P.; Klementová, M.; Krysiak, Y.; Palatinus, L. Accurate lattice parameters from 3D electron diffraction data. I. Optical distortions. *IUCrJ* **2022**, *9*, 735–755. [CrossRef] [PubMed]
16. Kresse, G.; Hafner, J. Ab initio molecular dynamics for liquid metals. *Phys. Rev. B* **1993**, *47*, 558. [CrossRef]
17. Sun, J.; Ruzsinszky, A.; Perdew, J. Strongly Constrained and Appropriately Normed Semilocal Density Functional. *Phys. Rev. Lett.* **2015**, *115*, 036402. [CrossRef]

18. Varignon, J.; Bibes, M.; Zunger, A. Mott gapping in 3d ABO_3 perovskites without Mott-Hubbard interelectronic repulsion energy U. *Phys. Rev. B* **2019**, *100*, 035119. [CrossRef]
19. Blöchl, P.E. Projector augmented-wave method. *Phys. Rev. B* **1994**, *50*, 17953. [CrossRef]
20. Sage, M.H.; Blake, G.R.; Marquina, C.; Palstra, T.T.M. Competing orbital ordering in RVO_3 compounds: High-resolution x-ray diffraction and thermal expansion. *Phys. Rev. B* **2007**, *76*, 195102. [CrossRef]
21. Miyasaka, S.; Okimoto, Y.; Iwama, M.; Tokura, Y. Spin-orbital phase diagram of perovskite-type RVO_3 (R = rare-earth ion or Y). *Phys. Rev. B* **2003**, *68*, 100406. [CrossRef]
22. Sage, M.H. Orbital, Charge and Magnetic order of RVO_3 Perovskites. Ph.D. Thesis, University of Groningen, Groningen, The Netherlands, 2006.

symmetry

MDPI

Article

Making the Most of 3D Electron Diffraction: Best Practices to Handle a New Tool

Khai-Nghi Truong [1,†], Sho Ito [2,3,†], Jakub M. Wojciechowski [1], Christian R. Göb [1], Christian J. Schürmann [1], Akihito Yamano [2], Mark Del Campo [4], Eiji Okunishi [5], Yoshitaka Aoyama [5], Tomohiro Mihira [5], Naoki Hosogi [5], Jordi Benet-Buchholz [6], Eduardo Carmelo Escudero-Adán [6], Fraser J. White [1], Joseph D. Ferrara [4] and Robert Bücker [1,*,†]

1 Rigaku Europe SE, Hugenottenallee 167, 63263 Neu-Isenburg, Germany
2 Rigaku Corporation, 3-9-12 Matsubara, Akishima 196-8666, Japan
3 DIC Central Research Laboratories, 631, Sakado, Sakura 285-8668, Japan
4 Rigaku Americas Corporation, 9009 New Trails Drive, The Woodlands, TX 77381, USA
5 JEOL Ltd., 3-1-2 Musashino, Akishima 196-8558, Japan
6 Institute of Chemical Research of Catalonia (ICIQ), The Barcelona Institute of Science and Technology, Av. Països Catalans 16, 43007 Tarragona, Spain
* Correspondence: robert.buecker@rigaku.com
† These authors contributed equally to this work.

Abstract: Along with the adoption of three-dimensional electron diffraction (3D ED/MicroED) as a mainstream tool for structure determination from sub-micron single crystals, questions about best practices regarding each step along the workflow, from data collection to structure solutions, arise. In this paper, we discuss three particular aspects of a 3D ED/MicroED experiment which, after hundreds of structures solved in Rigaku's laboratories, we have found to be important to consider carefully. First, for a representative model system of a hydrated compound (trehalose dihydrate), we show that cryo-transfer of the sample into the diffractometer is an effective means to prevent dehydration, while cooling of the sample without cryo-transfer yields a marginal improvement only. Next, we demonstrate for a small (tyrosine) and a large (clarithromycin) organic compound, how a simplified and fast workflow for dynamical diffraction calculations can determine absolute crystal structures with high confidence. Finally, we discuss considerations and trade-offs for choosing an optimal effective crystal-to-detector distance; while a long distance is mandatory for a protein (thaumatin) example, even a small molecule with difficult diffraction behavior (cystine) yields superior results at longer distances than the one used by default.

Keywords: electron diffraction; 3D ED; MicroED; single crystal diffraction; structure determination; absolute structure; cryo-transfer; crystallography; instrumentation

1. Introduction

Crystal structure determination using electron diffraction has seen an unprecedented revival starting from the late 2000s, sparked by the introduction of new data collection methods. These methods are analogous to single-crystal X-ray crystallography, wherein three-dimensional datasets are collected by stepwise or continuous rotation of single crystals in the probe beam without alignment to specific zone axes. These approaches are now commonly referred to under the umbrella terms 3D ED and MicroED [1–6]. The ability to obtain structures from single crystals in the sub-micron size range has generated huge interest in a wide range of scientific communities in organic and inorganic chemistry, as well as structural biology. Excellent review papers have recently become available, to which we would like to direct the reader for further introduction to the history, scope, and applications of 3D ED [6–9].

Hitherto, 3D ED has been performed in transmission electron microscopes (TEMs) which have typically been augmented by specific data collection software and/or hardware extensions such as precession units or dedicated diffraction detectors. More recently, dedicated electron diffractometers have become commercially available [10,11], which eliminate the requirement of prior skill in TEM operation and offer a user experience and workflows closely resembling those of single-crystal X-ray diffractometers. As such, 3D ED can be seamlessly integrated into the portfolio of crystallographic facilities. Despite their relatively recent introduction, dedicated electron diffractometers have already contributed significantly to published 3D ED studies from various fields [12–23]. Here, we specifically make use of the Rigaku XtaLAB Synergy-ED diffractometer [10], the central design principle of which is the complete integration of the collection, data processing, and structure solution process. The operational interface and user experience match that of Rigaku Oxford Diffraction X-ray diffractometers, driven by the fully integrated software package CrysAlis^{Pro} [24]. The electron optics of the XtaLAB Synergy-ED are based on a 200 kV electron gun with a lanthanum hexaboride crystal as the electron emitter. They were developed in collaboration with JEOL Ltd., based on JEOL's established platform for TEMs, specifically regarding the gun optics, the mechanical design, and the fully automated vacuum system. Notably, the goniometer of the XtaLAB Synergy-ED, despite some modifications to optimize performance for 3D ED, remains compatible with the wide portfolio of sample holders available for TEMs, such as cryo-transfer systems, as discussed below.

In this paper, based on lessons learned from more than 380 structures solved on XtaLAB Synergy-ED diffractometers in our application laboratories, we will discuss considerations for obtaining high-quality small-molecule and protein structures regarding sample cooling and cryo-transfer, radiation dose, crystal-to-detector distance, and absolute structure determination through dynamical refinement.

2. Materials and Methods

All small-molecule samples were commercially available, obtained in powder form, and used without further modification. Trehalose was purchased as both the dihydrate and anhydrous forms (>98% purity) from Tokyo Chemical Industry Germany GmbH (Eschborn, Germany). Clarithromycin (>99.5% purity) was purchased from LKT Laboratories (St. Paul, MN, USA), Inc. L-Tyrosine (>99% purity) was purchased from FUJIFILM Wako Pure Chemical Corporation (Osaka, Japan). Cystine was purchased from CARL ROTH ($\geq$98.5% purity, Karlsruhe, Germany).

2.1. Sample Grid Preparation

For small-molecule experiments, the microcrystalline powders were directly applied to continuous or lacey-carbon TEM grids after gentle grinding of sample powder between glass slides in order to break up larger crystals and agglomerates.

Thaumatin protein crystals were grown by batch crystallization from a 50 mg/mL solution in 100 mM ADA buffer (pH 6.5) by the addition of 1.5 M potassium sodium tartrate (pH 7.0) at a volume ratio of 1:1 at 20 °C. Next, 4 µL aliquots were applied to glow-discharged holey-carbon-coated copper grids at room temperature, blotted, and plunge-frozen into liquid ethane using a Leica GP2 plunger. One of the prepared sample grids was transferred to a Gatan Elsa (Model 698) cryo-transfer holder and introduced into a JEOL JIB-4000PLUS focused ion beam mill for further thinning [25]. Thin lamellae were cut out of 10 crystals using a gallium ion beam of successively lower currents (1 nA, 350 pA, 23 pA) down to approximately 250 nm thickness which is close to the optimal thickness for 3D ED [26,27]. The milling process took approximately 20 min per crystal on average. The grid was then directly cryo-transferred while remaining on the sample holder into the diffractometer for data collection.

2.2. Data Collection

All data collection was performed on identical Rigaku XtaLAB Synergy-ED diffractometers [10] located at various sites. Diffraction data are collected using an integrated Rigaku HyPix-ED hybrid-pixel detector (775 × 385 pixels, $(100\ \mu m)^2$ pixel size, frame rate up to 130 Hz) in zero dead time mode. Due to the event counting capability of the detector, readout and dark noise are not observed, ensuring the detectability of single electrons and no constraints in the choice of scan widths. The XtaLAB Synergy-ED diffractometer is fully controlled by the Rigaku Oxford Diffraction CrysAlis^{Pro} program package [24], which, in addition to real-space screening and data collection (instant and queued), also performs integrated on-the-fly and off-line data processing. Additionally, automatic structure solution and refinement for small molecules are possible within CrysAlis^{Pro} using AutoChem [28]. JEOL tomography-retainer and Gatan Elsa (Model 698) cryo-transfer sample holders were used for room-temperature and cryogenic measurements, respectively.

After insertion of a sample grid into the diffractometer, it is inspected in real space using the diffractometer's *visual mode*. First, a very low-resolution image covering the entire grid for coarse navigation is taken (*minimap*). Next, interactive screening is performed using *video-mode* or *snapshot-mode* visualization at a suitable magnification for identifying grains with morphological properties optimal for electron diffraction. For typical organic compounds or proteins as studied in this work, a crystal thickness of the order of 500 nm along the directions covered by the rotation scan is desirable for an optimal trade-off between signal strength and reasonable levels of background due to inelastic scattering. An extremely low electron flux (dose rate) of the order of $5 \times 10^{-4}\ e^-/(\text{Å}^2{\cdot}s)$ is used for screening to prevent radiation damage before data collection. Diffraction snapshots of candidates can be immediately collected, or grain positions can be queued for automatic screening, similar to existing workflows for 3D ED [29–31]. For each grain that is found to be well-diffracting, centering along the beam propagation direction is performed using a semi-automatic point-and-click interface. Afterward, data collection can either be immediately started or queued for unattended collection. All data collections are performed in shutterless continuous-rotation mode [4,6] at scan speeds of typically 0.5–3 s/° and fluxes of 1×10^{-3} to $5 \times 10^{-3}\ e^-/(\text{Å}^2{\cdot}s)$. A selected-area aperture of 1 μm or 2 μm apparent diameter (imaged on the sample plane) is used for isolation of the diffraction signal from the chosen grain. During data collection, unit cell determination and data reduction are continuously performed on-the-fly in CrysAlis^{Pro}, and online structure solution and refinement using AutoChem are started once data completeness passes a preset threshold.

For all small-molecule datasets, diffraction data were typically collected over wedges of 90–125° at total radiation doses (fluence) of 0.1–10 $e^-/\text{Å}^2$ per run (see Supplementary Crystallographic Tables for details). For thaumatin, diffraction data were collected over a total range of 70° from one of the crystals with a total dose of 10 $e^-/\text{Å}^2$.

2.3. Data Processing and Structure Solution

For all datasets, indexing, integration, and space-group determination were performed using CrysAlis^{Pro} 43.49a (L-tyrosine, clarithromycin) or 43.70a (thaumatin, trehalose, cystine). For small-molecule datasets, after dataset scaling and merging in CrysAlis^{Pro}, crystal structures were solved using direct methods and refined in the kinematical approximation in Olex2, using SHELXT [32] and SHELXL [33] as respective back-ends. For some datasets, the extinction parameter (EXTI) was used in refinement to prevent negative anisotropic displacement factors (ADPs) [34]. Further dynamical refinement using Jana2020 [35,36] was performed on some datasets, as will be discussed below. For thaumatin, after the export of unmerged data and subsequent merging using AIMLESS [37], molecular-replacement phasing was performed using PHASER [38]. Multiple rounds of model building with Coot 0.9.8 [39] and refinement with REFMAC5 [40] were performed until arriving at a completed model with satisfactory R-factors and geometry. Note that sum of Gaussian electron scattering factors was used in REFMAC5.

3. Results

In this section, we will discuss results from measurements on typical test samples that address three key aspects of experiment design unique to 3D ED: (i) the need to keep hydration/solvation of the specimen under vacuum; (ii) crystal structure refinement using dynamical diffraction calculations; and (iii) selection of appropriate crystal-to-detector distance using electron optics.

3.1. Preservation and Dehydration of Trehalose Dihydrate

As in X-ray crystallography, cooling the samples during an electron diffraction experiment [41] can significantly enhance data quality in terms of diffraction resolution and completeness through reduced thermal motion giving rise to an increase in scattering power. In electron diffraction, however, beam damage is generally a larger concern but can be mitigated to an extent with the help of low temperatures providing a longer lifetime of the crystal while in the beam. For a broad range of samples, in our laboratories, we often observe a similar behavior for electron diffraction, such that the time overhead for cooling the sample is more than offset by the faster screening for well-diffracting grains and lower requirements for careful dose optimization. A very different and even more crucial use case for cryo-preservation arises for samples such as hydrated and solvent-containing compounds, co-crystals, or proteins, where exposure to vacuum would lead to loss of hydration or solvent [17,23], often even leading to an entire collapse of the crystal lattice. For such samples, dehydration occurs quickly after transfer into the diffractometer, so simple cooling of the sample stage is not sufficient protection, and cryo-transfer, whereby the sample is frozen outside the vacuum and transferred into the diffractometer using a suitable cryo-transfer holder, is required.

As a representative organic molecule illustrating the typically observed behavior of hydrated small-molecule compounds, we studied the case of trehalose dihydrate, comparing results from ED experiments obtained at room temperature, at liquid-nitrogen temperature, and using cryo-transfer. We will compare the results to the case of natively anhydrous trehalose.

First, we introduced a sample grid of trehalose dihydrate into the diffractometer at room temperature. Among ~20 grains chosen for screening by their morphological properties (like grains shown in Figure 1g), most exhibited no diffraction or only very poor diffraction; only 3 grains were found suitable for data collection at an ultra-low total dose of 0.16 $e^-/Å^2$ and structure solution. In Figure 1b, a screening image, a representative diffraction frame, and the refined structure from one of those grains are shown. Full dehydration of the crystal due to exposure to the diffractometer vacuum was found for all grains. We then cooled down the sample to 100 K while keeping it in the vacuum environment of the diffractometer and repeated the measurement on the same grain to obtain a dataset with marginally improved statistics (Figure 1c, Table S1 in Supplementary Materials). Finally, we repeated the measurement with a total dose three times higher, which led to a further improvement in data refinement quality, specifically with regard to the signal-to-noise ratio (Table S1). For all datasets, we found an orthorhombic unit cell with volume $\approx$ 1450 $Å^3$ and a space group $P2_12_12_1$, matching a polymorph of anhydrous trehalose which has previously been described in the literature to arise from direct dehydration of trehalose dihydrate [42].

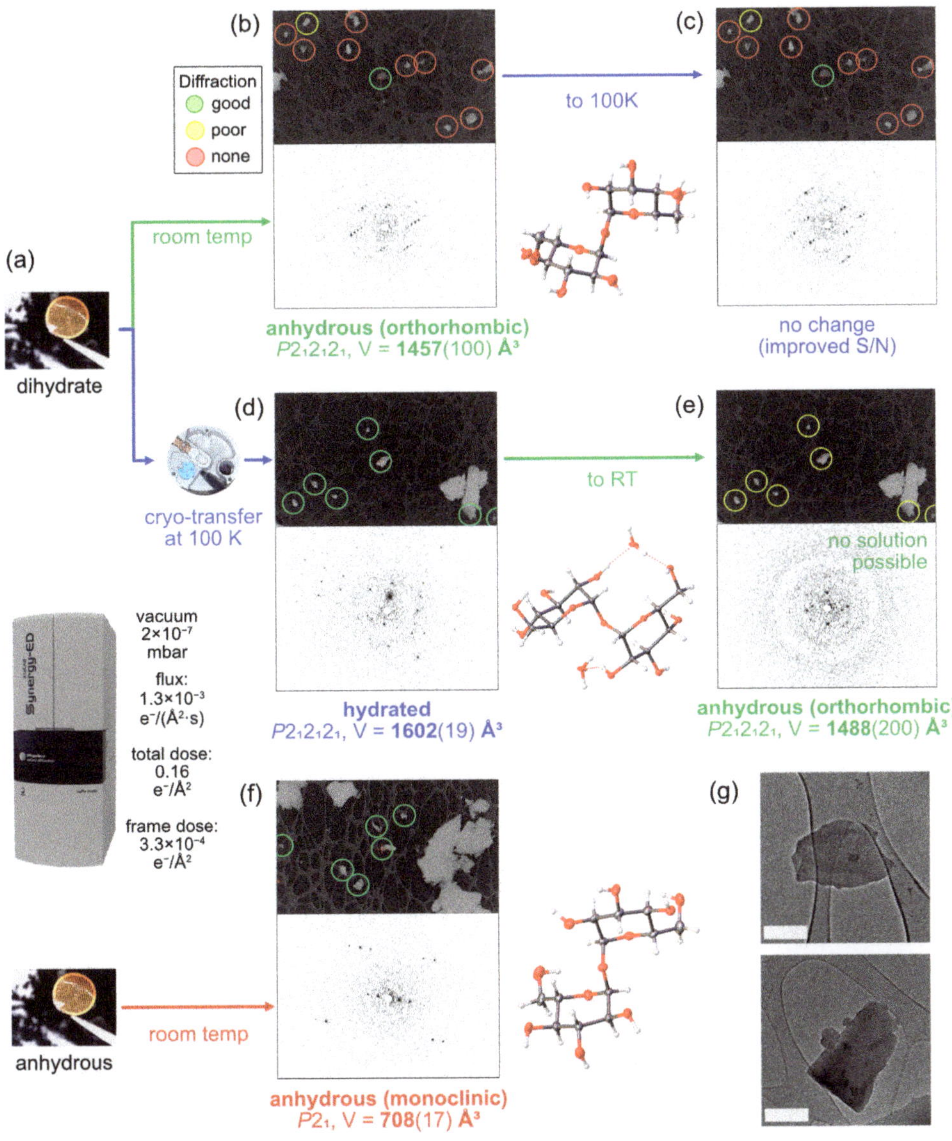

Figure 1. Dehydration and preservation of trehalose dihydrate inside the diffractometer. (**a**) Grids with powder of trehalose dihydrate (**top**) or anhydrous trehalose (**bottom**) are loaded on lacey-carbon coated copper grids and loaded into the diffractometer at either room temperature or at 100 K, using a cryo-transfer system; 3D ED datasets are taken at ultra-low dose to minimize the impact of radiation damage on our findings. (**b**) Typical screening result, diffraction pattern, and structure solution of anhydrous trehalose found after dehydration of the dihydrate in vacuum. Green, yellow, and red circles in the screening images indicate well-diffracting, poorly diffracting, and non-diffracting crystals, respectively. (**c**) Data from the same crystal in (**b**), after cooling to 100 K inside the vacuum. (**d**) Result for a grid with dihydrates preserved by cryo-transfer. (**e**) Data from the same crystal in (**d**), after warming up to room temperature inside the vacuum. (**f**) Result for a grid containing specifically prepared anhydrous trehalose, transferred into vacuum at room temperature. (**g**) Close-up real-space images of two typical grains taken after data collection. Scale bars correspond to 500 nm.

In the case of cryo-transferred trehalose dihydrate, we found good diffraction from all inspected morphologically suitable grains. The unit cell matched an orthorhombic cell with a slightly bigger volume of 1602(19) Å^3 than that of the anhydrous form, again in accordance with the literature [42,43]. We collected diffraction data from seven such grains at an ultra-low radiation dose of 0.16 $e^-/\text{Å}^2$ each, all of which led to a structure solution containing the two co-crystallized water molecules as shown in Figure 1d, unambiguously demonstrating the ability to preserve hydration. Due to the rather high-symmetry point group, a dataset of sufficient completeness could be collected from a single grain. After collecting the dihydrate crystal data, we warmed up the sample to ambient temperature while it remained in the vacuum environment of the diffractometer's airlock chamber. Revisiting the grains from which each data set was collected showed no visible morphological changes; however, only weak, low-resolution diffraction data could be obtained (Figure 1e). While data quality was too low for structure solution, we could determine the unit cells of the checked crystals; despite high uncertainties, the most likely value for all crystals was close to that of the previously found orthorhombic polymorph of the anhydrous state ($V \approx 1450$ Å^3), indicating dehydration of the crystals during warm-up in vacuum.

To compare the behavior of the vacuum-dehydrated trehalose to an initially anhydrous sample, we repeated a measurement at room temperature following the same protocol with commercially available anhydrous trehalose. Different from the case of dehydration inside the vacuum, we observed good diffraction for all screened crystals and could solve the structure for nine datasets like that shown in Figure 1f. Furthermore, those crystals belonged to the monoclinic polymorph, which, instead of dehydration, is obtained after crystallization at a temperature above the melting point of the dihydrate [44].

3.2. Dynamical Refinement Using CrysAlis^{Pro} and Jana2020

As of now, most structures in 3D ED are solved and refined using tools developed for X-ray diffraction, which besides adjusted atomic scattering factors, do not attempt to account for the properties of electron radiation. Specifically, the high elastic-scattering cross-sections of electrons lead to multiple diffraction and mutual interference between all excited beams (dynamical diffraction). The widely used assumption in crystallographic software that Bragg spot intensities should be proportional to the squared amplitude of the corresponding structure factors (kinematical diffraction) does not hold in this case, which is, however, used by most crystallographic refinement software. Making use of the kinematical approximation still yields satisfactory results, as data collection off zone axes [1], mosaicity on the nano-scale [45], and inelastic scattering [46] mitigate dynamical effects [9]. However, the usefulness of using full dynamical computations for refinement has now been demonstrated for a range of samples and techniques [34,47–50]. Specific advantages, besides improved refinement figures of merit such as R-values, include improved visibility of hydrogen atoms and guest molecules in framework materials thanks to lower noise in the reconstructed potentials [48,50], improved assignment of atom species [50], and sensitivity to the absolute structure of crystals [49] through differences between symmetry-equivalent reflections arising from dynamical effects. Dynamical refinement has been implemented in the structure solution and refinement package Jana2006/Jana2020 [35,36] for 3D ED data collected via the methods of precession tomography or, more recently, continuous rotation [50]. For the latter collection type, which is employed in the XtaLAB Synergy-ED diffractometer, unmerged intensities need to be computed for overlapping virtual frames (OVFs) at regular angular increments following the prescription of Klar et al. [50]. This currently can be performed using either the PETS2 processing software [51] or CrysAlis^{Pro} version 43.22a or higher. Due to the complexity and many parameters of the processing pipelines, differences between the final intensities assigned to the diffraction spots in the OVFs are expected between the processing programs. As a benchmark case, we collected 3D ED data from a nanocrystal of the amino acid *L*-tyrosine. After kinematical refinement without hydrogens and isotropic displacement parameters, we obtain a solution with R-values $R_1 = 14.4\%$ and $wR_2 = 35.4\%$. We next subjected this initial model to further dynamical refinement in Jana2020 following the procedures as outlined in the literature [36,50]. In consecutive steps, we added (1) frame scaling and crystal thickness (with fixed atomic structure), (2) the atom positions and isotropic displacement parameters, (3) free hydrogen positions, and (4) anisotropic displacement parameters (ADPs) as refinement parameters. This procedure was repeated for data processed by CrysAlis^{Pro} 43.49a and PETS2.2, respectively, with the initial model as *L*- and inverted *D*-enantiomer (equivalently referred to as *S*- and *R*-enantiomer). The resulting R-values are displayed in Figure 2b,c. Regardless of the input data and refinement type, we observe a clear preference for the correct chirality, and each of the refinement steps further decreases the R-values until they finally reach approximately half their values from kinematical refinement (R_1 = 7.6%, wR_2 = 22% for CrysAlis^{Pro} and *L*-configuration, Table S2), in accordance with typical behavior documented in the literature. There are minor quantitative differences between refinements against OVF intensities obtained from PETS2 and CrysAlis^{Pro} (CAP); remarkably, while CrysAlis^{Pro}-processed data refined to lower R_1 values, the wR_2 values are higher, which we attributed to differences in the error model corrections of the programs. Still, we find no major qualitative deviations. To obtain a more quantitative measure for the confidence level of absolute structure assignment, as originally suggested by Le Page et al. [52], we compute the background noise-adjusted z-score for the bias of observed intensities towards those computed from the correct absolute configuration following the definition of Klar et al. [50]. Z-scores higher than 3σ were found for even the simplest computation, with values reaching up to 10.0σ for data processing in CAP and full refinement in Jana2020, as shown in Figure 2d. Also, here, we find only minor deviations between the processing programs, likely arising from noise models and details of OVF computation.

Another representative example of absolute structure determination using CrysAlis^{Pro}/ Jana2020 for a larger chiral compound is the structure of the common antibiotic drug clarithromycin. A structure solution from a grain found to be of the solvent-free form II polymorph [53,54] is shown in Figure 2e, which after kinematical refinement in SHELXL resulting in R_1 = 14.8%/wR_2 = 31.9%, was subjected to dynamical refinement in Jana2020 with crystal thickness (EDthick), atom positions, and isotropic displacement parameters as free parameters, leading to a large gap between the correct (R_1 = 13.0%/wR_2 = 25.7%, Table S3) and incorrect (R_1 = 16.7%/wR_2 = 31.3%) absolute configuration with an adjusted z-score of 14.0.

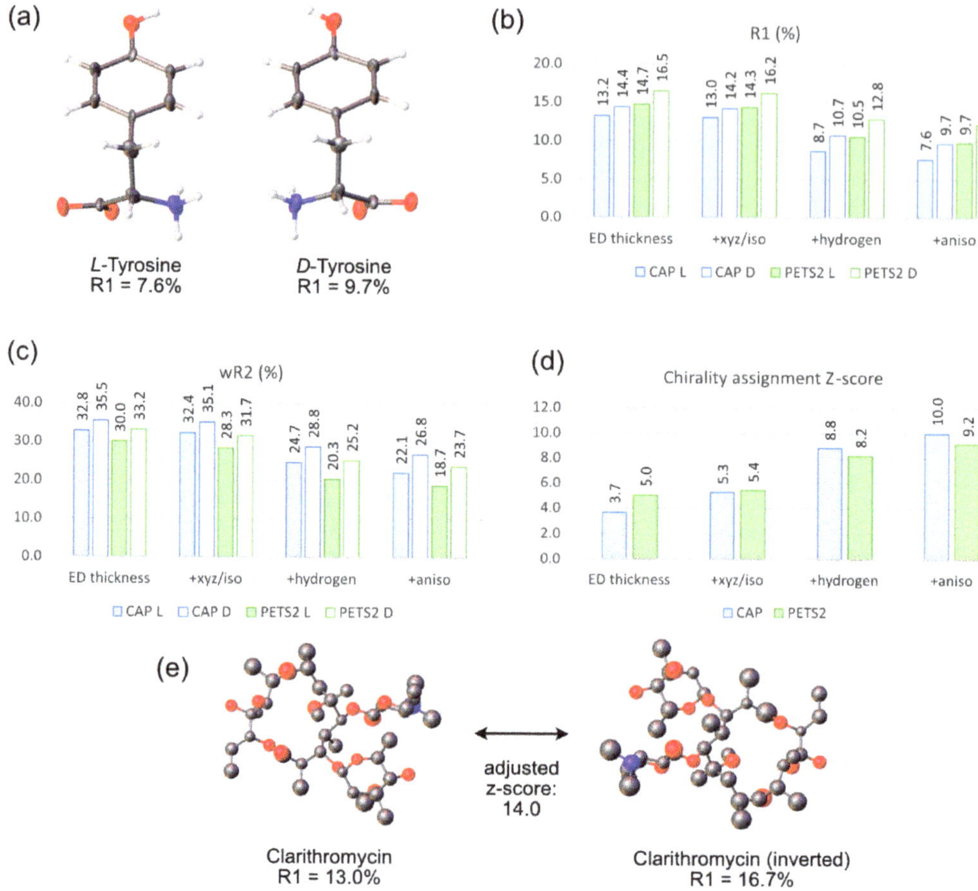

Figure 2. Dynamical refinement of crystal structures from L-tyrosine and clarithromycin. (**a**) Drawings of the tyrosine structure solved as *L*- and *D*-enantiomer with resulting R_1 values after dynamical refinement. (**b**) R_1 values after dynamical refinement of tyrosine crystal structures processed with CrysAlis^{Pro} (CAP, blue) and PETS2 (green) for *L*- (filled) and *D*- (empty) enantiomers after several types of refinements applied. (**c**) As (**b**), with wR_2 as figure of merit. (**d**) Z-score (background-noise-adjusted) for correct determination of enantiomer by comparing observed intensities to predictions from refinements with both *L*- and *D*-configuration. (**e**) Correct and inverted configuration of clarithromycin after dynamical refinement with respective R_1 value.

3.3. Operation at Large Crystal-to-Detector Distances for Proteins and Small Molecules

All TEMs (except for some dedicated STEMs) allow one to set the magnification of the projection of images onto the detector by means of changing currents in the electron lenses below the sample. If set to diffraction mode, this corresponds to changing the effective crystal-to-detector distance (DD), equivalent to changing the physical detector position along the beam direction on an X-ray diffractometer. Featuring a similar projection system, the XtaLAB Synergy-ED diffractometers used in this study allow one to choose between calibrated DD settings ranging between approximately 500 mm and 2500 mm, providing the ability to both collect data at high resolutions up to ≈0.3 Å, as well as resolve closely spaced reflections arising from the large unit cells of, among others, protein crystals or mesoporous compounds.

For the former scenario, the question arises of what minimum DD and hence peak spacing on the newly introduced detector is required for a typical protein crystal when using the combination of the HyPix-ED detector and processing in CrysAlis$^{Pro.}$ We collected a dataset from a focused ion beam milled lamella of a thaumatin microcrystal (Figure 3a,b) with a tetragonal unit cell (a = b = 58.4 Å, c = 151.2 Å), which is a commonly used standard for 3D ED on proteins [55,56]. DD was set to 1800 mm, potentially keeping spots out to ~1.2 Å (resolution at detector edge) covered by the detector. In Figure 3c,d, we show frames from the dataset viewed perpendicular to the *c*-axis with a Bragg spot spacing of 0.30 mm (3 pixels) on the detector. Despite this small spacing, the data could be integrated by CrysAlis^{Pro} using default two-pass profile fitting and were further processed using a standard molecular-replacement pipeline to a final protein structure at a resolution of 2.15 Å (Figure 3e, see Table S4 and Section 2.3 for details).

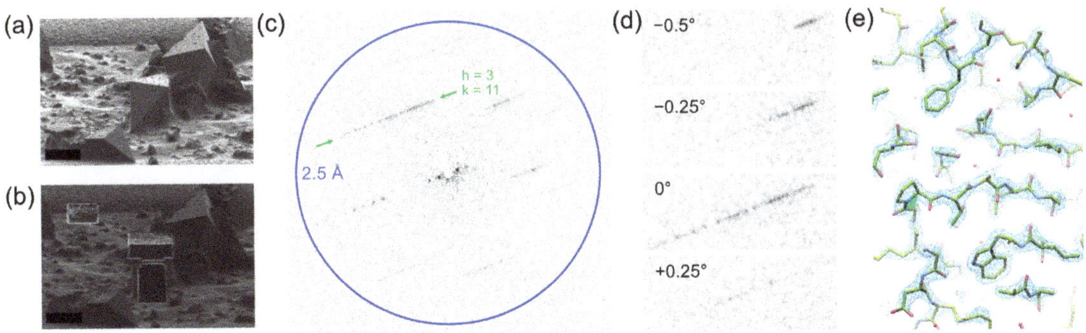

Figure 3. Three-dimensional ED data collection from a protein microcrystal. (**a**,**b**) SEM images of thaumatin crystals on a vitrified TEM grid imaged before (**a**) and after (**b**) FIB milling. Scale bars correspond to 10 μm. The thin lamella section in (**b**) is used for data collection. (**c**) Diffraction frame from continuous rotation collection over a wedge of 0.25° nearly perpendicular to the tetragonal c-axis. The indicated streak corresponds to reflections with h = 3, k = 11. (**d**) Close-ups of the indicated streak from the frame shown in (**c**) and adjacent frames spaced by 0.25°. (**e**) Section from structure solution with $2F_{obs} - F_{calc}$ at 2σ.

While the advantages of using a short DD for small-molecule crystallography, in particular on inorganic compounds with small unit cells and diffracting to high resolutions, are obvious, the case for longer DD is more subtle. For a challenging test sample exhibiting an unusually long unit cell axis as well as a tendency for twinning and a poor overall diffraction signal, we used cystine, an oxidized amino acid forming a hexagonal polymorph with a *c*-axis of ≈58 Å length. Both at 100 K and 293 K, we found most crystal grains to be very poorly diffracting or badly twinned. Using automated screening collection (unattended queue mode, 4° collection range) from approximately 100 manually selected candidate crystals on two grids, we identified two well-diffracting, only weakly intergrown crystals at starkly different orientations with respect to the sample grid (Figure 4a–f). While crystal 1 exhibited a plate-like morphology and was oriented with its long *c*-axis almost perpendicular to the grid (Figure 4a–c), crystal 2 had its *c*-axis almost in the plane of the sample grid (Figure 4d–f). For each of the crystals, we conducted three data collections at an identical crystal orientation and scan range but different settings of DD. We also used a fine frame slicing/scan width of 0.15° for optimal resolution of peaks in three dimensions, which, due to the noise-free readout of the hybrid-pixel detector, does not impact signal-to-noise ratios.

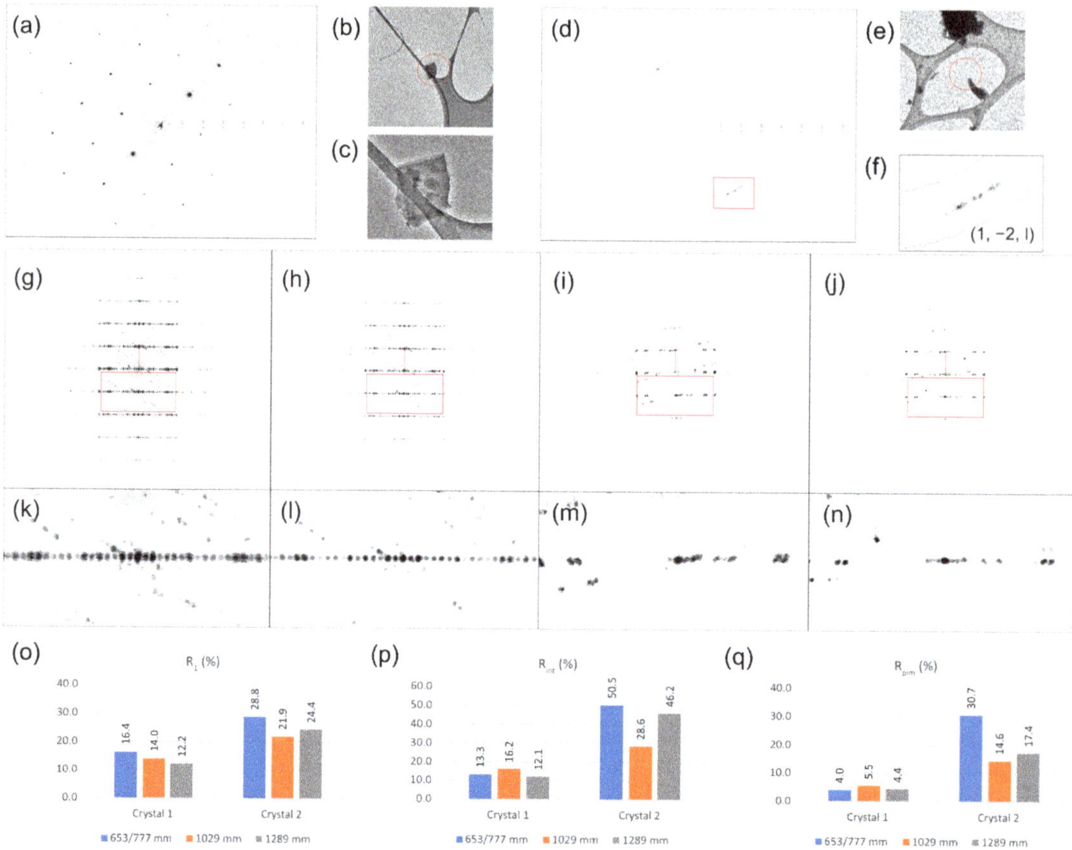

Figure 4. Diffraction data collection from two grains of hexagonal cystine. (**a**) Diffraction pattern near neutral tilt at DD = 1289 mm, (**b**) real-space image, (**c**) close-up image for crystal 1. The red circle in (**b**) corresponds to the region that diffraction was collected from using a selected-area aperture of apparent diameter 2 μm. (**d**) Diffraction pattern near neutral tilt at DD = 1289 mm, with red box indicating zoom region in (**f**), (**e**) real-space image, (**f**) zoom into diffraction pattern for crystal 2. (**g**) 3D reconstruction of reciprocal space from pixel intensities, projected along b* for crystal 1 at DD = 653 mm. The shown region corresponds to $(0.8\ \text{Å})^{-1}$ along each axis. The red box indicates the zoom region shown in (**k**). (**h**) Same, at DD = 1289 mm. (**i**) Same, for crystal 2 at DD = 777 mm. (**j**) Same, for crystal 2 at DD = 1289 mm. (**k–n**) Zoomed sections of (**g–j**), respectively, highlighting reflections with h = 1. (**o**) R_1 values after refinement in SHELXL, for three DD values for each crystal. (**p**) R_{int} values after integration, scaling, and merging in CrysAlis^{Pro}. (**q**) R_{pim} values.

Overall, crystal 1 yielded significantly better data, as expected, due to the absence of strong peak overlaps with the *c**-axis never passing the (almost flat) Ewald sphere within the collected wedge of 125°. Additionally, the thin plate-like morphology minimizes the detrimental effects of inelastic scattering by diffusing the spots and dynamical diffraction. In Figure 4g,h, projections of reciprocal space reconstructed from pixel intensities in the collected frames are shown along the *b**-axis for the shortest and longest DD. While peaks are well-separated and aligned in an undistorted lattice along all axes, the long DD yields better-defined spots along the *a**-direction, as seen in the close-up views of peaks with h = 1 in Figure 4k,l. This comes at the expense of a more pronounced cut-off at higher resolution due to more limited detector coverage. Despite the unproblematic geometry, we observed the best refinement metric R_1 from the longest DD of 1289 mm (Figure 4o,

Table S5) at similar precision indicators R_{int} and R_{pim} (Figure 4p,q). This suggests that the higher data redundancy achieved at low DD is outweighed by higher accuracy of integrated intensities due to better peak fitting and decreased overlap, specifically with a weak intergrown second crystal.

The case of crystal 2 is significantly more challenging, as frames collected almost normal to the grid exhibit tightly spaced and more diffuse peaks (Figure 4d,f). The latter is due to the crystal, despite its small size, retaining a plate-like geometry and hence a greater thickness perpendicular to the *c*-axis. As can be seen in Figure 4i,j,m,n, besides a lower resolution (cut at 1.1 Å for processing), the reconstructed reciprocal space shows poorly separated peaks as well as spatial distortions, specifically at the lowest DD. This was chosen as 777 mm instead of 653 mm, as for crystal 1, to avoid peak spacing along c^* falling below 0.3 mm. We attribute this to a poor determination of crystal orientation. Both issues are mitigated by longer DD settings, significantly improving the quality of processing and refinement (Figure 4o–q, Table S5). Optimal quality is reached at the intermediate distance, which we hypothesize to be related to spurious factors such as the onset of radiation damage or crystal drift in the longest-DD dataset.

4. Discussion

With the advent of commercially available, fully integrated diffractometers [10,11], 3D ED is transitioning from an exploratory new approach developed within a dedicated method-centered community to a cutting-edge technique offered by an increasing number of crystallographic facilities. As such, it is becoming available to a wide audience of non-specialist users from diverse fields who might not yet be familiar with practicalities that differ from common knowledge in X-ray crystallography. In this work, we conducted systematic measurements on test systems, specifically targeting three such aspects to be considered for a successful 3D ED experiment. We will further discuss the outcomes of those in the following.

4.1. Cryo-Transfer Allows Data Collection from Hydrates and Solvates

The ability to protect samples from dehydration in the diffractometer vacuum is critical for a wide range of substances, such as hydrates, solvates, and biomolecules. While the creation of atmospheric micro-environments [57–59] using specific sample holders is possible, cryo-transfer, where the sample grid is cooled to low temperature outside the vacuum and transferred into the vacuum in a cold state, is a significantly simpler, faster, and cheaper workflow. While cryo-transfer is now commonplace in cryo-electron microscopy or 3D-ED/MicroED on biomolecules [7,60], its strict necessity for many small molecule samples is less appreciated.

Our findings on dehydration of trehalose dihydrate (and prevention thereof) in Section 3.1 are entirely consistent with the hypothesis of complete preservation of crystallites using cryo-transfer. Insertion into a vacuum without cryo-protection or warm-up after cryo-transfer not only leads to dehydration, but in most cases, also partial or complete loss of crystallinity. As a corollary, this means that there may be cases where it is not possible to distinguish between a sample that has collapsed due to being unprotected in a vacuum and one that exhibits poor crystallinity from the start. This underlines the necessity of additional PXRD measurements, ideally during or after exposure to vacuum, if no cryo-transfer is possible. Furthermore, we could show that in-vacuum cooling of the small fraction of crystals that dehydrated without collapse of the crystal lattice only leads to a marginal improvement in data quality. It was recently reported by Yang et al. [61] that cryogenic data collection even decreased data quality for test systems, including an anhydrous sugar, highlighting that the impact of temperature for vacuum-insensitive samples is highly sample-dependent and warrants further study. Of note, dehydrated crystals were of the rare orthorhombic polymorph [42], which differs in its unit cell parameters from the dihydrate by a small but well-detectable margin, whereas commercially obtained anhydrous crystals were also measured and found to be well-diffracting and of the more common monoclinic polymorph [44].

While trehalose dihydrate is only one specific example of a hydrated microcrystalline compound, the observed behaviors are commonly found in our laboratories for a wide range of small molecule samples, specifically pharmaceuticals and solvated mesoporous compounds, which are regularly solved on request by commercial and academical customers and collaborators.

4.2. Absolute Structure Determination Using a Simplified Workflow

Among the virtues of dynamical diffraction effects in electron diffraction, the determination of absolute structure and, thus, the chirality of the studied compound [49,50] is particularly outstanding for the field of pharmaceuticals and for organic small molecules, among others. The necessary intensity differences between Bijvoet-pairs, however, arise from a fundamentally different physical mechanism than that of resonance scattering of X-rays and cannot be accounted for within the kinematical approximation, such that only dynamical refinement strategies based on Bloch-wave or multi-slice computations [35,62–64] are able to make use of the chiral information contained in 3D ED data.

Inevitably, routine application of dynamical refinement for absolute structure determination poses new challenges both in terms of knowledge and practical skills of crystallographers, as well as the required computation time of the order of hours for a full dynamical refinement, even on high-end workstations. While progress in available packages [36] will help to somewhat lower those barriers, the necessary workflows are still significantly more intricate than what typical users of highly integrated X-ray diffractometers are used to.

As we have shown by a benchmarking series on a standard sample in Section 3.2, data processing using the latest versions of CrysAlis^{Pro} provides data suitable for dynamical refinement (OVF intensities). The data output leads to results comparable to that achieved by reprocessing in PETS2 [51] when used for dynamical refinement in Jana2020 [36,65]. OVF intensities are computed automatically during data collection on an XtaLAB Synergy-ED diffractometer, such that, along with a starting model provided by AutoChem/Olex2 [28], a full set of input files for dynamical refinement is available immediately after data collection. This significantly speeds up the workflow for dynamical refinement. Furthermore, we showed that even without the refinement of atomic coordinates and ADPs, comparison to dynamically predicted intensities after frame scaling and crystal thickness refinement [65] gives a satisfactory degree of confidence for absolute structure determination, as gauged by z-scores well exceeding 3σ, and differences in R-values shown in Figure 2. For determination of absolute structure with the highest fidelity, as well as full exploitation of the information contained in 3D ED data from both organic and inorganic compounds, such as high visibility of hydrogens and guest molecules, more accurate bond lengths, and more confident assignment of atom species [50], a full refinement (presented in detail elsewhere [50,65]) is inevitably required; however, for proper absolute structure determination, a single round of dynamical parameter optimization in Jana2020 following CrysAlis^{Pro} data reduction can give a good indication of absolute structure within a few minutes after data collection. Very recently, the absolute structures of the active pharmaceutical ingredient levocetirizine dihydrochloride and the natural product hakuhybotrol were determined using similar workflows [16,20].

4.3. Appropriate Detector Distance Can Improve Data Quality

TEMs, as well as the XtaLAB Synergy-ED diffractometer, allow setting the detector distance prior to measurement by merely changing a preset in the control software, leading to questions on how to pick an optimal detector distance (DD) for a given sample. For protein crystals, long DDs are clearly required at the expense of high-resolution coverage, which scales reciprocally with the accessible unit cell length. As a rule of thumb, the DD should be larger than $(p\Delta N) \cdot c/\lambda$, where p, c, λ denote pixel size, longest unit cell axis, and wavelength, respectively. The factor ΔN, representing the number of pixels that peaks should be separated by, depends on the point-spread function of the detector, as well as the performance of the used integration algorithm. The successful data processing of a thaumatin protein crystal collected at a DD set to an aggressively low value of ≈1800 mm indicates that $\Delta N \approx 3$ is sufficient for the combination of the HyPix-ED detector with CrysAlis^{Pro}. We attribute this to the relatively large pixel size of 100 μm leading to a narrow point spread even with 200 keV electrons [66] and the sophisticated multi-pass profile fitting protocols in CrysAlis^{Pro}. This finding implies that a maximum unit cell length of more than 200 Å is still compatible with standard settings on the XtaLAB Synergy-ED, even while retaining detector coverage up to 1.7 Å at near the far edge of the detector.

Besides the case of protein crystals, small-molecule 3D ED can also profit from longer DD settings, as we have demonstrated by the example of cystine nanocrystals, combining an exceptionally long unit cell axis, a tendency for intergrowth and twinning, and a high symmetry. For the case of a crystal showing relatively low-resolution diffraction and with a large portion of the collected data having close peak spacing on each frame due to the flat Ewald sphere lying close to a long axis and diffuse inelastic scattering, common settings of DD for small-molecule work were unable to yield satisfactory data quality. It is evident that increasing DD in this scenario leads to a significant improvement. More surprisingly, long DD settings otherwise associated with protein crystals even slightly improved data quality in an unproblematic crystal where no close peak spacing on single frames was observed, but the reduced data redundancy at high resolutions might have been outweighed by decreased residual peak overlap for rejection of a small intergrown crystal, and higher accuracy of crystal orientation and peak fitting. It should be noted that this trade-off might be more unfavorable in the case of lower symmetry, where low completeness at high resolutions due to angular coverage becomes a limiting factor. Also, while our measurements have been conducted on a counting detector with no background noise, an instrument fitted with a conventional CMOS or CCD detector might yield worse signal-to-noise ratios at larger DD as more background noise is collected from the larger integration regions required to fit the magnified peaks. Still, those findings indicate that, for data collection of compounds with unknown properties with regards to diffraction resolution, propensity for intergrowth and twinning, as well as diffuse inelastic scattering, the choice of a larger DD than usual should be taken into consideration, despite a possible penalty in completeness or precision (though not accuracy) of reflection intensities. This might be especially useful for upcoming applications in high-throughput, automated screening of heterogeneous mixtures [67–70], which will frequently contain such sub-optimal crystals that nevertheless need to be included in data analysis as effectively as possible. At the other end of the spectrum, inorganic compounds will often yield useful diffraction up to very high resolutions; in this case, further shortening the DD below typical values will lead to improved resolution and completeness, which would be useful, e.g., for charge-density studies or dynamical refinement.

5. Conclusions and Outlook

In this paper, we conducted measurements on representative samples, addressing three particular aspects of experiment design for 3D ED/MicroED, that is, preservation of hydration by cryo-transfer, simple dynamical refinement for determination of absolute structure, and optimal choice of effective crystal-to-detector distance. All those are of immediate interest for practical work on TEMs and electron diffractometers, using workflows and instrumentation available in dedicated research groups and, increasingly, general crystallographic facilities.

Along with the adoption of 3D ED/MicroED into the canon of structure determination methods, new approaches regarding sample preparation, data collection, data processing, and structure solution are rapidly developing, such as high-throughput crystallinity screening [67], on-grid crystallization [71], lipid cubic phase embedded crystals [72], environmental cells [57,58,73], efficient analysis of heterogeneous samples [68–70], autonomous and serial data collection [69,74–76], low-resolution phasing [77], dynamical refinement [47,48,64], absolute structure determination [49,50], diffuse scattering and pair-distribution functions [78,79], and aspherical atom models [80], among many others. Assessing the new opportunities provided by such exciting developments in everyday work and casting them into dependable workflows will be an important task for years to come. Hence, while our current study is limited to relatively basic parameters, we are hoping for many more investigations on deriving best practices for obtaining the best structures possible, from a broad array of sample types, while minimizing the effort for individual optimization for each sample.

Supplementary Materials: The following supporting information can be downloaded at: https://www.mdpi.com/article/10.3390/sym15081555/s1, Table S1: Trehalose; Table S2: Tyrosine; Table S3: Clarithromycin; Table S4: Thaumatin; Table S5: Cystine.

Author Contributions: Sample preparation and data collection were performed by K.-N.T., S.I., J.M.W., C.R.G., C.J.S., A.Y., E.O., Y.A., T.M., N.H. and R.B. Data processing and structure solution were performed by K.-N.T., S.I., M.D.C., J.B.-B., E.C.E.-A. and R.B. Crystallographic data files were prepared by K.-N.T., M.D.C. and R.B. The project was conceived and super-vised by A.Y., F.J.W., J.D.F. and R.B. The manuscript was written by R.B. with contributions from all authors. All authors have read and agreed to the published version of the manuscript.

Funding: This research received no external funding.

Data Availability Statement: Several molecular structures from this work are deposited in the Cambridge Structural Database (CSD) or, respectively, the Protein Database (PDB), using accession codes as quoted in the crystallographic data tables. Ancillary molecular structures solved solely for the purpose of systematic comparisons are available upon request from the authors.

Acknowledgments: We are grateful to Hiroyasu Sato, Takashi Kikuchi, and Keigo Takahira for assistance with experiment data management. We appreciate helpful discussions on data processing with Mathias Meyer Michał Jasnowski and Przemysław Stec.

Conflicts of Interest: All authors except for J.B.-B. and E.C.E.-A. are current or former employees of subsidiaries of the Rigaku Holdings Corporation or JEOL Limited, which are partners in the development of the XtaLAB Synergy-ED diffractometer. They have no conflicting interests, financial or otherwise, and declare that their affiliation with either company had no influence on the design or interpretation of the experiments.

References

1. Kolb, U.; Gorelik, T.; Kübel, C.; Otten, M.T.; Hubert, D. Towards Automated Diffraction Tomography: Part I-Data Acquisition. *Ultramicroscopy* **2007**, *107*, 507–513. [CrossRef] [PubMed]
2. Mugnaioli, E.; Gorelik, T.; Kolb, U. "Ab Initio" Structure Solution from Electron Diffraction Data Obtained by a Combination of Automated Diffraction Tomography and Precession Technique. *Ultramicroscopy* **2009**, *109*, 758–765. [CrossRef] [PubMed]
3. Zhang, D.; Oleynikov, P.; Hovmöller, S.; Zou, X. Collecting 3D Electron Diffraction Data by the Rotation Method. *Z. Krist.* **2010**, *225*, 94–102. [CrossRef]
4. Nederlof, I.; van Genderen, E.; Li, Y.-W.; Abrahams, J.P. A Medipix Quantum Area Detector Allows Rotation Electron Diffraction Data Collection from Submicrometre Three-Dimensional Protein Crystals. *Acta Crystallogr. D Biol. Crystallogr.* **2013**, *69*, 1223–1230. [CrossRef]
5. Shi, D.; Nannenga, B.L.; Iadanza, M.G.; Gonen, T. Three-Dimensional Electron Crystallography of Protein Microcrystals. *eLife* **2013**, *2*, e01345. [CrossRef]
6. Gemmi, M.; Mugnaioli, E.; Gorelik, T.E.; Kolb, U.; Palatinus, L.; Boullay, P.; Hovmöller, S.; Abrahams, J.P. 3D Electron Diffraction: The Nanocrystallography Revolution. *ACS Cent. Sci.* **2019**, *5*, 1315–1329. [CrossRef]
7. Nannenga, B.L.; Gonen, T. The Cryo-EM Method Microcrystal Electron Diffraction (MicroED). *Nat. Methods* **2019**, *16*, 369–379. [CrossRef]
8. Gruene, T.; Holstein, J.J.; Clever, G.H.; Keppler, B. Establishing Electron Diffraction in Chemical Crystallography. *Nat. Rev. Chem.* **2021**, *5*, 660–668. [CrossRef]
9. Saha, A.; Nia, S.S.; Rodríguez, J.A. Electron Diffraction of 3D Molecular Crystals. *Chem. Rev.* **2022**, *122*, 13883–13914. [CrossRef]
10. Ito, S.; White, F.J.; Okunishi, E.; Aoyama, Y.; Yamano, A.; Sato, H.; Ferrara, J.D.; Jasnowski, M.; Meyer, M. Structure Determination of Small Molecule Compounds by an Electron Diffractometer for 3D ED/MicroED. *CrystEngComm* **2021**, *23*, 8622–8630. [CrossRef]
11. Simoncic, P.; Romeijn, E.; Hovestreydt, E.; Steinfeld, G.; Santiso-Quiñones, G.; Merkelbach, J. Electron Crystallography and Dedicated Electron-Diffraction Instrumentation. *Acta Crystallogr. Sect. E Crystallogr. Commun.* **2023**, *79*, 410–422. [CrossRef] [PubMed]

12. Kahlenberg, V.; Vinke, J.; Krüger, H.; Ito, S.; Schürmann, C.J. $Na_2Ca_3Si_2O_8$ or γ-$Na_2Ca_6Si_4O_{15}$? A Hybrid Approach Combining 3D Single-Crystal Electron and Powder X-Ray Diffraction. *J. Am. Ceram. Soc.* **2022**, *105*, 6976–6998. [CrossRef]
13. Pearce, N.; Reynolds, K.E.A.; Kayal, S.; Sun, X.Z.; Davies, E.S.; Malagreca, F.; Schürmann, C.J.; Ito, S.; Yamano, A.; Argent, S.P.; et al. Selective Photoinduced Charge Separation in Perylenediimide-Pillar[5]Arene Rotaxanes. *Nat. Commun.* **2022**, *13*, 415. [CrossRef] [PubMed]
14. Steinke, F.; Otto, T.; Ito, S.; Wöhlbrandt, S.; Stock, N. Isostructural Family of Rare-Earth MOFs Synthesized from 1,1,2,2-Tetrakis(4-Phosphonophenyl)Ethylene. *Eur. J. Inorg. Chem.* **2022**, *2022*, e202200562. [CrossRef]
15. Jandl, C.; Steinfeld, G.; Li, K.; Pang, P.K.C.; Choi, C.L.; Wang, C.; Simoncic, P.; Williams, I.D. Absolute Structure Determination of Chiral Zinc Tartrate MOFs by 3D Electron Diffraction. *Symmetry* **2023**, *15*, 983. [CrossRef]
16. Karothu, D.P.; Alhaddad, Z.; Göb, C.R.; Schürmann, C.J.; Bücker, R.; Naumov, P. The Elusive Structure of Levocetirizine Dihydrochloride Determined by Electron Diffraction. *Angew. Chem. Int. Ed.* **2023**, *62*, e202303761. [CrossRef]
17. Mies, T.; Schürmann, C.; Ito, S.; White, A.J.P.; Crimmin, M.R.; Barrett, A.G.M. Synthesis and Characterization of a Calcium-Pyrazolonato Complex. Observation of In-Situ Desolvation during Micro-Electron Diffraction. *Z. Anorg. Allg. Chem.* **2023**, *649*, e202200294. [CrossRef]
18. Perera, T.A.; Taylor, W.V.; Gildner, M.B.; Reinheimer, E.W.; Ito, S.; Nelson, A.; Yost, S.R.; Hudnall, T.W. Photochemical Reactions of a Diamidocarbene: Cyclopropanation of Bromonaphthalene, Addition to Pyridine, and Activation of Sp^3 C–H Bonds. *Chem. Sci.* **2023**, *14*, 7867–7874. [CrossRef]
19. Shivanna, M.; Zheng, J.-J.; Ray, K.G.; Ito, S.; Ashitani, H.; Kubota, Y.; Kawaguchi, S.; Stavila, V.; Yao, M.-S.; Fujikawa, T.; et al. Selective Sorption of Oxygen and Nitrous Oxide by an Electron Donor-Incorporated Flexible Coordination Network. *Commun. Chem.* **2023**, *6*, 62. [CrossRef]
20. Watanabe, Y.; Takahashi, S.; Ito, S.; Tokiwa, T.; Noguchi, Y.; Azami, H.; Kojima, H.; Higo, M.; Ban, S.; Nagai, K.; et al. Hakuhybotrol, a Polyketide Produced by Hypomyces Pseudocorticiicola, Characterized with the Assistance of 3D ED/MicroED. *Org. Biomol. Chem.* **2023**, *21*, 2320. [CrossRef]
21. Xue, Z.; Zheng, J.-J.; Nishiyama, Y.; Yao, M.-S.; Aoyama, Y.; Fan, Z.; Wang, P.; Kajiwara, T.; Kubota, Y.; Horike, S.; et al. Fine Pore-Structure Engineering by Ligand Conformational Control of Naphthalene Diimide-Based Semiconducting Porous Coordination Polymers for Efficient Chemiresistive Gas Sensing. *Angew. Chem. Int. Ed.* **2023**, *62*, e202215234. [CrossRef]
22. Andersen, C.E.; McPherson, J.N.; Giménez-Marqués, M.; Li, J.; Kubus, M.; Ito, S.; Goeb, C.R.; Ott, S.; Larsen, R.W.; Espallargas, G.M.; et al. Vapor-Phase Synthesis of Low-Valent Metal-Organic Frameworks from Metal Carbonyl Synthons. *J. Mater. Chem. C* **2023**. [CrossRef]
23. Kubens, L.; Truong, K.-N.; Lehmann, C.W.; Lützenkirchen-Hecht, D.; Bornhorst, J.; Mohr, F. The structure of Maneb: An important manganese-containing bis(dithiocarbamate) fungicide. *Chem. Eur. J.* **2023**, e202301721. [CrossRef] [PubMed]
24. *Rigaku Oxford Diffraction CrysAlisPro Software System*; Rigaku Corporation: Wrocław, Poland, 2023.
25. Duyvesteyn, H.M.E.; Kotecha, A.; Ginn, H.M.; Hecksel, C.W.; Beale, E.V.; de Haas, F.; Evans, G.; Zhang, P.; Chiu, W.; Stuart, D.I. Machining Protein Microcrystals for Structure Determination by Electron Diffraction. *Proc. Natl. Acad. Sci. USA* **2018**, *115*, 9569–9573. [CrossRef] [PubMed]
26. Peet, M.J.; Henderson, R.; Russo, C.J. The Energy Dependence of Contrast and Damage in Electron Cryomicroscopy of Biological Molecules. *Ultramicroscopy* **2019**, *203*, 125–131. [CrossRef]
27. Martynowycz, M.W.; Clabbers, M.T.B.; Unge, J.; Hattne, J.; Gonen, T. Benchmarking the Ideal Sample Thickness in Cryo-EM. *Proc. Natl. Acad. Sci. USA* **2021**, *118*, e2108884118. [CrossRef]
28. *Rigaku Oxford Diffraction AutoChem in Conjunction with Olex2*; Rigaku Corporation: Wrocław, Poland, 2023.
29. Cichocka, M.O.; Ångström, J.; Wang, B.; Zou, X.; Smeets, S. High-Throughput Continuous Rotation Electron Diffraction Data Acquisition via Software Automation. *J. Appl. Crystallogr.* **2018**, *51*, 1652–1661. [CrossRef]
30. de la Cruz, M.J.; Martynowycz, M.W.; Hattne, J.; Gonen, T. MicroED Data Collection with SerialEM. *Ultramicroscopy* **2019**, *201*, 77–80. [CrossRef]
31. Cheng, A.; Negro, C.; Bruhn, J.F.; Rice, W.J.; Dallakyan, S.; Eng, E.T.; Waterman, D.G.; Potter, C.S.; Carragher, B. Leginon: New Features and Applications. *Protein Sci.* **2021**, *30*, 136–150. [CrossRef]
32. Sheldrick, G.M. SHELXT—Integrated Space-Group and Crystal-Structure Determination. *Acta Crystallogr. Sect. Found. Adv.* **2015**, *71*, 3–8. [CrossRef]
33. Sheldrick, G.M. A Short History of SHELX. *Acta Crystallogr. A* **2008**, *64*, 112–122. [CrossRef] [PubMed]
34. Yang, T.; Xu, H.; Zou, X. Improving Data Quality for Three-Dimensional Electron Diffraction by a Post-Column Energy Filter and a New Crystal Tracking Method. *J. Appl. Crystallogr.* **2022**, *55*, 1583–1591. [CrossRef]
35. Palatinus, L.; Petříček, V.; Corrêa, C.A. Structure Refinement Using Precession Electron Diffraction Tomography and Dynamical Diffraction: Theory and Implementation. *Acta Crystallogr. Sect. Found. Adv.* **2015**, *71*, 235–244. [CrossRef] [PubMed]
36. Petříček, V.; Palatinus, L.; Plášil, J.; Dušek, M. JANA2020—A New Version of the Crystallographic Computing System Jana. *Z. Krist. Cryst. Mater.* **2023**. [CrossRef]
37. Evans, P.R.; Murshudov, G.N. How Good Are My Data and What Is the Resolution? *Acta Crystallogr. D Biol. Crystallogr.* **2013**, *69*, 1204–1214. [CrossRef] [PubMed]
38. McCoy, A.J.; Grosse-Kunstleve, R.W.; Adams, P.D.; Winn, M.D.; Storoni, L.C.; Read, R.J. Phaser Crystallographic Software. *J. Appl. Crystallogr.* **2007**, *40*, 658–674. [CrossRef]
39. Emsley, P.; Lohkamp, B.; Scott, W.G.; Cowtan, K. Features and Development of Coot. *Acta Crystallogr. D Biol. Crystallogr.* **2010**, *66*, 486–501. [CrossRef]

40. Murshudov, G.N.; Skubák, P.; Lebedev, A.A.; Pannu, N.S.; Steiner, R.A.; Nicholls, R.A.; Winn, M.D.; Long, F.; Vagin, A.A. REFMAC5 for the Refinement of Macromolecular Crystal Structures. *Acta Crystallogr. D Biol. Crystallogr.* **2011**, *67*, 355–367. [CrossRef]
41. Denysenko, D.; Grzywa, M.; Tonigold, M.; Streppel, B.; Krkljus, I.; Hirscher, M.; Mugnaioli, E.; Kolb, U.; Hanss, J.; Volkmer, D. Elucidating Gating Effects for Hydrogen Sorption in MFU-4-Type Triazolate-Based Metal-Organic Frameworks Featuring Different Pore Sizes. *Chem. Eur. J.* **2011**, *17*, 1837–1848. [CrossRef]
42. Nagase, H.; Ogawa, N.; Endo, T.; Shiro, M.; Ueda, H.; Sakurai, M. Crystal Structure of an Anhydrous Form of Trehalose: Structure of Water Channels of Trehalose Polymorphism. *J. Phys. Chem. B* **2008**, *112*, 9105–9111. [CrossRef]
43. Brown, G.M.; Rohrer, D.C.; Berking, B.; Beevers, C.A.; Gould, R.O.; Simpson, R. The Crystal Structure of α,α-Trehalose Dihydrate from Three Independent X-Ray Determinations. *Acta Crystallogr. B* **1972**, *28*, 3145–3158. [CrossRef]
44. Jeffrey, G.A.; Nanni, R. The Crystal Structure of Anhydrous α,α-Trehalose at -150°. *Carbohydr. Res.* **1985**, *137*, 21–30. [CrossRef] [PubMed]
45. Gallagher-Jones, M.; Ophus, C.; Bustillo, K.C.; Boyer, D.R.; Panova, O.; Glynn, C.; Zee, C.; Ciston, J.; Mancia, K.C.; Minor, A.M.; et al. Nanoscale Mosaicity Revealed in Peptide Microcrystals by Scanning Electron Nanodiffraction. *Commun. Biol.* **2019**, *2*, 26. [CrossRef] [PubMed]
46. Latychevskaia, T.; Abrahams, J.P. Inelastic Scattering and Solvent Scattering Reduce Dynamical Diffraction in Biological Crystals. *Acta Crystallogr. Sect. B Struct. Sci. Cryst. Eng. Mater.* **2019**, *75*, 523–531. [CrossRef] [PubMed]
47. Palatinus, L.; Corrêa, C.A.; Steciuk, G.; Jacob, D.; Roussel, P.; Boullay, P.; Klementová, M.; Gemmi, M.; Kopeček, J.; Domeneghetti, M.C.; et al. Structure Refinement Using Precession Electron Diffraction Tomography and Dynamical Diffraction: Tests on Experimental Data. *Acta Crystallogr. Sect. B* **2015**, *71*, 740–751. [CrossRef]
48. Palatinus, L.; Brázda, P.; Boullay, P.; Perez, O.; Klementová, M.; Petit, S.; Eigner, V.; Zaarour, M.; Mintova, S. Hydrogen Positions in Single Nanocrystals Revealed by Electron Diffraction. *Science* **2017**, *355*, 166–169. [CrossRef]
49. Brázda, P.; Palatinus, L.; Babor, M. Electron Diffraction Determines Molecular Absolute Configuration in a Pharmaceutical Nanocrystal. *Science* **2019**, *364*, 667–669. [CrossRef]
50. Klar, P.B.; Krysiak, Y.; Xu, H.; Steciuk, G.; Cho, J.; Zou, X.; Palatinus, L. Accurate Structure Models and Absolute Configuration Determination Using Dynamical Effects in Continuous-Rotation 3D Electron Diffraction Data. *Nat. Chem.* **2023**, *15*, 848–855. [CrossRef]
51. Palatinus, L.; Brázda, P.; Jelínek, M.; Hrdá, J.; Steciuk, G.; Klementová, M. Specifics of the Data Processing of Precession Electron Diffraction Tomography Data and Their Implementation in the Program PETS2.0. *Acta Crystallogr. Sect. B Struct. Sci. Cryst. Eng. Mater.* **2019**, *75*, 512–522. [CrossRef]
52. Le Page, Y.; Gabe, E.J.; Gainsford, G.J. A Robust Alternative to η Refinement for Assessing the Hand of Chiral Compounds. *J. Appl. Crystallogr.* **1990**, *23*, 406–411. [CrossRef]
53. Iwasaki, H.; Sugawara, Y.; Adachi, T.; Morimoto, S.; Watanabe, Y. Structure of 6-O-Methylerythromycin A (Clarithromycin). *Acta Crystallogr. C* **1993**, *49*, 1227–1230. [CrossRef]
54. Tian, J.; Thallapally, P.K.; Dalgarno, S.J.; Atwood, J.L. Free Transport of Water and CO2 in Nonporous Hydrophobic Clarithromycin Form II Crystals. *J. Am. Chem. Soc.* **2009**, *131*, 13216–13217. [CrossRef]
55. de la Cruz, M.J.; Hattne, J.; Shi, D.; Seidler, P.; Rodriguez, J.; Reyes, F.E.; Sawaya, M.R.; Cascio, D.; Weiss, S.C.; Kim, S.K.; et al. Atomic-Resolution Structures from Fragmented Protein Crystals with the CryoEM Method MicroED. *Nat. Methods* **2017**, *14*, 399–402. [CrossRef] [PubMed]
56. Blum, T.B.; Housset, D.; Clabbers, M.T.B.; van Genderen, E.; Bacia-Verloop, M.; Zander, U.; McCarthy, A.A.; Schoehn, G.; Ling, W.L.; Abrahams, J.P. Statistically Correcting Dynamical Electron Scattering Improves the Refinement of Protein Nanocrystals, Including Charge Refinement of Coordinated Metals. *Acta Crystallogr. Sect. Struct. Biol.* **2021**, *77*, 75–85. [CrossRef]
57. Karakulina, O.M.; Demortière, A.; Dachraoui, W.; Abakumov, A.M.; Hadermann, J. In Situ Electron Diffraction Tomography Using a Liquid-Electrochemical Transmission Electron Microscopy Cell for Crystal Structure Determination of Cathode Materials for Li-Ion Batteries. *Nano Lett.* **2018**, *18*, 6286–6291. [CrossRef]
58. Batuk, M.; Vandemeulebroucke, D.; Ceretti, M.; Paulus, W.; Hadermann, J. Topotactic Redox Cycling in $SrFeO_{2.5+\delta}$ Explored by 3D Electron Diffraction in Different Gas Atmospheres. *J. Mater. Chem. A* **2022**, *11*, 213–220. [CrossRef]
59. Wu, S.; Li, J.; Ling, Y.; Sun, T.; Fan, Y.; Yu, J.; Terasaki, O.; Ma, Y. In Situ Three-Dimensional Electron Diffraction for Probing Structural Transformations of Single Nanocrystals. *Chem. Mater.* **2022**, *34*, 8119–8126. [CrossRef]
60. Clabbers, M.T.B.; Xu, H. Microcrystal Electron Diffraction in Macromolecular and Pharmaceutical Structure Determination. *Drug Discov. Today Technol.* **2020**, *37*, 93–105. [CrossRef]
61. Yang, T.; Waitschat, S.; Inge, A.K.; Stock, N.; Zou, X.; Xu, H. A Comparison of Structure Determination of Small Organic Molecules by 3D Electron Diffraction at Cryogenic and Room Temperature. *Symmetry* **2021**, *13*, 2131. [CrossRef]
62. Kirkland, E.J. Computation in Electron Microscopy. *Acta Crystallogr. Sect. Found. Adv.* **2016**, *72*, 1–27. [CrossRef]
63. Seto, Y.; Ohtsuka, M. *ReciPro*: Free and Open-Source Multipurpose Crystallographic Software Integrating a Crystal Model Database and Viewer, Diffraction and Microscopy Simulators, and Diffraction Data Analysis Tools. *J. Appl. Crystallogr.* **2022**, *55*, 397–410. [CrossRef]
64. Cleverley, A.; Beanland, R. Modelling Fine-Sliced Three Dimensional Electron Diffraction Data with Dynamical Bloch-Wave Simulations. *IUCrJ* **2023**, *10*, 118–130. [CrossRef] [PubMed]

65. Dušek, M.; Petříček, V.; Palatinus, L.; Rohlíček, J.; Fejfarová, K.; Eigner, V.; Kučeráková, M.; Poupon, M.; Henriques, M.; Plášil, J.; et al. *Jana2006 Cookbook*; Institute of Physics ASCR, v.v.i.: Prague, Czech Republic, 2019.
66. McMullan, G.; Cattermole, D.M.; Chen, S.; Henderson, R.; Llopart, X.; Summerfield, C.; Tlustos, L.; Faruqi, A. Electron Imaging with Medipix2 Hybrid Pixel Detector. *Ultramicroscopy* **2007**, *107*, 401–413. [CrossRef] [PubMed]
67. Delgadillo, D.; Burch, J.; Kim, L.J.; de Moraes, L.; Niwa, K.; Williams, J.; Tang, M.; Lavallo, V.; Chhetri, B.; Jones, C.; et al. High-Throughput Identification of Crystalline Natural Products from Crude Extracts Enabled by Microarray Technology and MicroED. *chemRxiv* **2023**. [CrossRef]
68. Lightowler, M.; Li, S.; Ou, X.; Cho, J.; Li, A.; Hofer, G.; Xu, J.; Yang, T.; Zou, X.; Lu, M.; et al. Phase Identification and Discovery of Hidden Crystal Forms in a Polycrystalline Pharmaceutical Sample Using High-Throughput 3D Electron Diffraction. *chemRxiv* **2023**. [CrossRef]
69. Luo, Y.; Wang, B.; Smeets, S.; Sun, J.; Yang, W.; Zou, X. High-Throughput Phase Elucidation of Polycrystalline Materials Using Serial Rotation Electron Diffraction. *Nat. Chem.* **2023**, *15*, 483–490. [CrossRef]
70. Unge, J.; Lin, J.; Weaver, S.J.; Her, A.S.; Gonen, T. Autonomous MicroED Data Collection Enables Compositional Analysis. *ChemRxiv* **2023**. [CrossRef]
71. Gillman, C.; Nicolas, W.J.; Martynowycz, M.W.; Gonen, T. Design and Implementation of Suspended Drop Crystallization. *IUCrJ* **2023**, *10*, 430–436. [CrossRef] [PubMed]
72. Martynowycz, M.W.; Shiriaeva, A.; Clabbers, M.T.B.; Nicolas, W.J.; Weaver, S.J.; Hattne, J.; Gonen, T. A Robust Approach for MicroED Sample Preparation of Lipidic Cubic Phase Embedded Membrane Protein Crystals. *Nat. Commun.* **2023**, *14*, 1086. [CrossRef]
73. Ling, Y.; Sun, T.; Guo, L.; Si, X.; Jiang, Y.; Zhang, Q.; Chen, Z.; Terasaki, O.; Ma, Y. Atomic-Level Structural Responsiveness to Environmental Conditions from 3D Electron Diffraction. *Nat. Commun.* **2022**, *13*, 6625. [CrossRef]
74. Smeets, S.; Zou, X.; Wan, W. Serial Electron Crystallography for Structure Determination and Phase Analysis of Nanocrystalline Materials. *J. Appl. Crystallogr.* **2018**, *51*, 1262–1273. [CrossRef] [PubMed]
75. Wang, B.; Zou, X.; Smeets, S. Automated Serial Rotation Electron Diffraction Combined with Cluster Analysis: An Efficient Multi-Crystal Workflow for Structure Determination. *IUCrJ* **2019**, *6*, 854–867. [CrossRef] [PubMed]
76. Bücker, R.; Hogan-Lamarre, P.; Mehrabi, P.; Schulz, E.C.; Bultema, L.A.; Gevorkov, Y.; Brehm, W.; Yefanov, O.; Oberthür, D.; Kassier, G.H.; et al. Serial Protein Crystallography in an Electron Microscope. *Nat. Commun.* **2020**, *11*, 996. [CrossRef] [PubMed]
77. Gorelik, T.E.; Lukat, P.; Kleeberg, C.; Blankenfeldt, W.; Müller, R. Molecular Replacement for Small Molecule Crystal Structure Determination from Electron Diffraction Data with Reduced Resolution. *Acta Cryst. A* **2023**. *submitted*.
78. Gorelik, T.E.; Neder, R.; Terban, M.W.; Lee, Z.; Mu, X.; Jung, C.; Jacob, T.; Kaiser, U. Towards Quantitative Treatment of Electron Pair Distribution Function. *Acta Crystallogr. Sect. B Struct. Sci. Cryst. Eng. Mater.* **2019**, *75*, 532–549. [CrossRef]
79. Poppe, R.; Vandemeulebroucke, D.; Neder, R.B.; Hadermann, J. Quantitative Analysis of Diffuse Electron Scattering in the Lithium-Ion Battery Cathode Material $Li_{1.2}Ni_{0.13}Mn_{0.54}Co_{0.13}O_2$. *IUCrJ* **2022**, *9*, 695–704. [CrossRef]
80. Gruza, B.; Chodkiewicz, M.L.; Krzeszczakowska, J.; Dominiak, P.M. Refinement of Organic Crystal Structures with Multipolar Electron Scattering Factors. *Acta Crystallogr. Sect. Found. Adv.* **2020**, *76*, 92–109. [CrossRef]

symmetry

MDPI

Article

Characterisation of Microstructure and Special Grain Boundaries in LPBF AlSi10Mg Alloy Subjected to the KoBo Extrusion Process

Przemysław Snopiński [1,*] and Krzysztof Matus [2]

[1] Department of Engineering Materials and Biomaterials, Silesian University of Technology, 18A Konarskiego Street, 44-100 Gliwice, Poland

[2] Materials Research Laboratory, Silesian University of Technology, 18A Konarskiego Street, 44-100 Gliwice, Poland; krzysztof.matus@polsl.pl

* Correspondence: przemyslaw.snopinski@polsl.pl

Abstract: Grain boundary engineering (GBE) enhances the properties of metals by incorporating specific grain boundaries, such as twin boundaries (TB). However, applying conventional GBE to parts produced through additive manufacturing (AM) poses challenges, since it necessitates thermomechanical processing, which is not desirable for near-net-shape parts. This study explores an alternative GBE approach for post-processing bulk additively manufactured aluminium samples (KoBo extrusion), which allows thermo-mechanical treatment in a single operation. The present work was conducted to examine the microstructure evolution and grain boundary character in an additively manufactured AlSi10Mg alloy. Microstructural evolution and grain boundary character were investigated using Electron Back Scattered Diffraction (EBSD) and Transmission Electron Microscopy (TEM). The results show that along with grain refinement, the fraction of Coincidence Site Lattice boundaries was also increased in KoBo post-processed samples. The low-Σ twin boundaries were found to be the most common Coincidence Site Lattice boundaries. On the basis of EBSD analysis, it has been proven that the formation of CSL boundaries is directly related to a dynamic recrystallisation process. The findings show prospects for the possibility of engineering the special grain boundary networks in AM Al–Si alloys, via the KoBo extrusion method. Our results provide the groundwork for devising GBE strategies to produce novel high-performance aluminium alloys.

Keywords: CSL boundaries; aluminium; KoBo extrusion; EBSD; TKD; TEM; texture

1. Introduction

Nature has always been full of symmetry and asymmetry. Many natural objects have symmetric shapes, which highlight the inherent order and balance in biological and geological systems. The symmetrical forms [1,2] such as the radial symmetry of a daisy or the bilateral symmetry of an orchid, exemplify the beauty of nature. The hexagonal symmetry of snowflakes [3,4], the bilateral symmetry of butterfly wings [5], and the symmetry of spiral galaxies [6] all demonstrate basic principles of symmetry in natural systems.

Symmetry also plays an important role in the field of materials science [7], where it provides a powerful framework for understanding the organisation, behaviour, and properties of materials at the atomic and molecular levels. In the context of metals science, symmetry manifests itself as a fundamental principle that underlies the arrangement of atoms and the formation of crystal structures [8]. The specific symmetry groups exhibited by the crystals provide insight into their physical properties, including mechanical behaviour, electrical conductivity, and optical response.

The characteristics of polycrystalline metals are closely related to the atomic subtleties associated with grain boundaries (GBs), which are fundamental structural elements and represent the interfaces where crystallographic orientations change abruptly. GBs could exhibit unique structural, chemical, and electronic properties that differ from those of bulk crystalline regions [9]. These characteristics have an important impact on various properties of the material, such as mechanical strength, corrosion resistance, and electrical conductivity, which is why grain boundaries are the subject of intense scientific investigation [10].

In pursuit of enhancing the properties of polycrystalline materials, researchers have taken two different approaches. One involves the engineering of material textures, while the other involves targeting special grain boundary distributions within the polycrystal. The motivation for the latter technique comes from the observation that different atomic structures at the grain boundary often give rise to different properties or responses [11]. Hence, the current essence of "grain boundary engineering" is to produce alloys with a high proportion of $\Sigma 3$ twin boundaries (TB) and twin-related higher-order $\Sigma 3n$ ($n = 2, 3$; $\Sigma \leq 29$) [12–14], referred to as low-sigma Coincidence Site Lattice (CSL) boundaries. As twin structures [15], including TBs, twins, and stacking faults (SFs), not only provide improved strength but also provide good ductility, harnessing the potential of CSL grain boundaries through precise manipulation and control therefore holds great promise for tailoring the properties of many polycrystalline materials [16].

Currently, the grain boundary engineering (GBE) of face-centred cubic (FCC) metallic materials, such as copper, nickel, lead alloys, and austenitic stainless steels, with low stacking fault energies (SFEs, primarily relies on thermo-mechanical processes (TMP). These processes involve cold deformation followed by heat treatment, taking advantage of the low SFE to promote the nucleation of annealing twins (ATs) in significant quantities. TMP can be categorised into two main approaches: strain-annealing and strain-recrystallisation. Both methods aim to increase the population of low-Σ (Coincidence Site Lattice) special grain boundaries (SBs) in metallic materials, which in turn modify random high-angle grain boundaries (RHAGBs) and hinder the connectivity of RHAGB networks [17]. However, materials with high SFEs face challenges in the development of twin structures or SBs. Consequently, the grain boundary engineering in aluminium and its alloys remains a complex undertaking [18].

To encourage the formation of growth twins in metals with high stacking fault energies (SFEs), a commonly employed strategy is to incorporate alloying elements. This incorporation results in the reduction of their SFEs, which facilitates the twinning process [19]. In recent years, researchers have also employed other techniques to successfully fabricate nanotwinned (NT) aluminium alloys. For example, surface engineering [20], cryogenic temperature deformation [21], and the incorporation of reinforcements such as carbon nanotubes (CNTs) [22,23] have been explored. Furthermore, bulk NT metals have also been successfully manufactured using high-rate, high-strain plastic deformation methods [24,25].

The KoBo extrusion process is a plastic deformation method that offers high-strain deformation at low temperatures [26], along with other plastic deformation techniques such as equal-channel angular pressing (ECAP) [27–29], high-pressure torsion (HPT) [30], and accumulative roll bonding (ARB) [31]. One distinguishing characteristic of KoBo extrusion is its ability to frequently change the deformation route, resulting in a substantial concentration of point defects at the atomic level, including vacancies and self-interstitial atoms (SIA), which alter the thermodynamic state of the material [32]. This characteristic is particularly intriguing for the fabrication of nanotwinned materials (NT), as the accumulation of SIAs has been associated with the occurrence of nanotwins [33,34]. Nevertheless, it remains unknown whether the KoBo extrusion process can induce the formation of high-density twins in laser powder bed fusion (LPBF) aluminium alloys, as this aspect has yet to be investigated. Furthermore, it is possible to unlock novel properties related to corrosion resistance, embrittlement, and fracture resistance by engineering the grain boundary network in Al-Si alloys of LPBF. This is because the crystallography of a grain boundary influences its diffusivity, mobility, and segregation behaviour. However, detailed knowledge of the crystallographic characteristics of Coincidence Site Lattice (CSL) grain boundaries in AM microstructures is currently lacking. In addition, the scope of materials engineering is expanding beyond traditional polycrystalline structures, with increasing interest in the synthesis and characterisation of interface-dominated structural alloys, such as nanostructured alloys.

The aim of this study is to explore an alternative grain boundary engineering (GBE) approach, specifically the KoBo extrusion post-processing method, for additively manufactured aluminium samples. The study investigates the evolution of the microstructure and grain boundary character of the AlSi10Mg alloy to assess the potential for engineering special grain boundary networks in additive manufacturing (AM) Al–Si alloys, ultimately paving the way for the design of AM-compatible GBE strategies to produce novel high-performance materials.

2. Materials and Methodology

The specimens were fabricated using the SLM TruPrint 1000 system from Trumpf (Ditzingen, Germany) using spherical powder of an AlSi10Mg alloy, Figure 1. The chemical composition of the powder used in this study, supplied by Sigma-Aldrich (Steinheim, Germany), is given in Table 1. The following SLM process parameters were used to fabricate dense samples: a laser power of 175 W, a layer thickness of 0.02 mm, a scanning speed of 1.4 m/s, and a zig-zag scanning strategy with an angle of 67 degrees rotation on each layer. Using these parameters, cylindrical specimens with a diameter of 60 mm and a height of 50 mm were successfully produced.

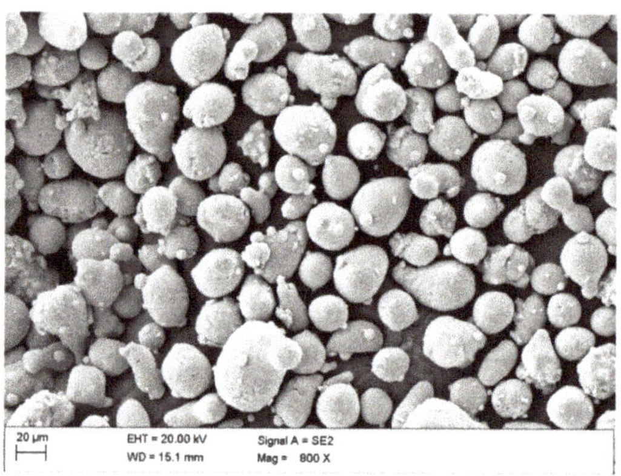

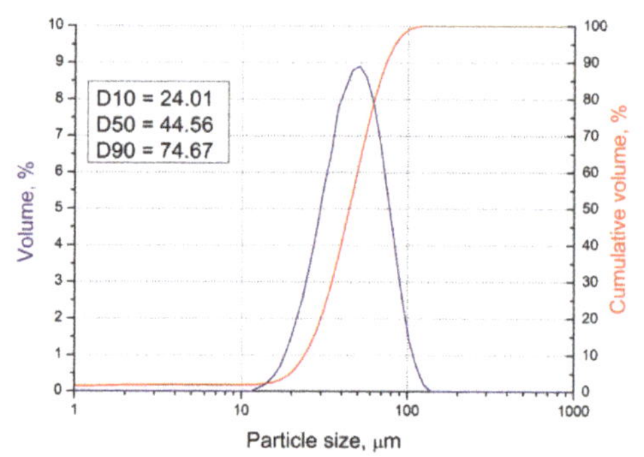

Figure 1. SEM image of the AlSi10Mg alloy powder used in this study and corresponding particle size distribution measurement.

Table 1. Chemical composition of the AlSi10Mg alloy powder.

Al	Mg	Si	Ti	Cu	Fe
87.8	0.5	10.5	0.15	0.15	0.09

The KoBo extrusion process, the principle of which is described in the referenced study [35], was carried out without sample preheating, and the extrusion process was carried out using the following parameters:

- punch speed of 0.2 mm/s;
- die rotation angle of $\pm 8^\circ$;
- frequency of 5 Hz;
- extrusion ratio λ of 225. The extrusion ratio λ signifies that the diameter of the rods after the KoBo extrusion process was equal to = 4 mm;
- maximal measured temperature close to the extrusion die (~280 °C);

- sample cooling—room-temperature water.

To conduct the microstructural investigation, a sample was carefully extracted from the middle part of the extruded rod. This specific location was chosen because it represents a region where the process parameters had stabilised.

A combination of light microscopy (LM), scanning electron microscopy (SEM), and transmission electron microscopy (TEM) was employed to study the microstructure in detail.

For LM investigations, the specimens were prepared using standard metallographic procedures. This included grinding (with 500 and 1200 grit SiC papers) and polishing (with 6, 3 and 1 μm diamond pastes) the samples to achieve a suitable mirror-like surface finish. The metallographic samples were etched with Keller's reagent (2.5 mL HNO_3, 1.5 mL HCl, 1 mL HF, and 95 mL H_2O) for 60 s.

A scanning electron microscope was used to study the microstructure at higher magnification. The sample preparation for SEM followed the same procedure as for LM, with additional fine polishing with 0.04 μm colloidal silica for 1 h. This additional polishing step was performed to reduce the roughness of the surface and thus improve the quality of the EBSD signal.

Electron Backscatter Diffraction (EBSD) analysis was conducted using a scanning electron microscope (Zeiss Supra 35, Carl Zeiss NTS GmbH, Oberkochen, Germany) equipped with an EDAX NT EBSD detector (38 mm diameter) controlled by TSL OIM software. The TSL OIM EBSD software generated a comprehensive data set and visualisations, offering detailed information on the crystallographic orientation and microstructure of the analysed material.

The surfaces perpendicular to the scanning (LPBF sample) and transverse direction (KoBo sample) were selected for the EBSD characterisation. The following EBSD process parameters were used:

- accelerating voltage: 20 kV;
- sample tilt: 72°;
- step size: 0.15 μm.

The EBSD orientation maps were then post-processed using ATEX software (Université de Lorraine: Metz, France) [36], which provided comprehensive information on grain boundary characteristics.

The grain-orientation-spread (GOS) maps were calculated based on the spread of crystallographic orientations within individual grains. The spread was measured in terms of misorientation angles between neighbouring measurement points within each grain. For a grain numbered i, its grain orientation spread, GOS(i), can be expressed as follows:

$$GOS(i) = \frac{1}{n_1} \sum_{j=1}^{i=n_i} \omega(g_j \langle g_i \rangle) \quad (1)$$

For the detection of recrystallised grains, a (GOS(i) < 2.5°) value was selected.

TEM analysis was performed to obtain detailed insights into the microstructure at the nanoscale. We used a Titan 80-300 FEI S/TEM microscope, which was operated at 300 kV. Electron diffraction images were further analysed using digital micrograph and CrysTBox (Crystallographic Toolbox) software. To fabricate the TEM sample, typical lamellae were cut from the extruded specimen using focused ion beam (FIB) milling. The lamellae were extracted parallel to the extrusion direction.

The TEM lamella was additionally analysed in SEM microscope using Transmission Kikuchi Diffraction (TKD) to obtain crystallographic information that allowed characterisation of grain boundaries with improved resolution. The sample was mounted in a custom-made holder that held it at a net angle of 10° with reference to the horizontal. The working distance was set at 7.5 mm. The energy of the electron beam was kept constant at 20 kV throughout the analysis. The TKD map was acquired in ~2 h with a step size of 20 nm and a TKD pattern resolution of 301 × 129 pixels. A total area of 12 μm × 5.2 μm was recorded. The TKD orientation map was then post-processed using ATEX software, which provided comprehensive information on grain boundary features. The CSL grain boundaries were evaluated according to the Brandon criterion of $\Delta\theta \leq 15°\Sigma^{-1/2}$ standard. To determine the sizes of grains, automated mean linear intercept measurement was employed.

3. Results

Figure 2 shows the microstructures of the AlSi10Mg alloy observed using a light microscope. In Figure 2a, one can see distinct and discontinuous laser scan traces. These scan tracks overlap with each other at an angle of approximately 67 degrees, which corresponds to the Zig-Zag scanning strategy used during the additive manufacturing process. This tracks arrangement is a characteristic feature of the LPBF microstructures.

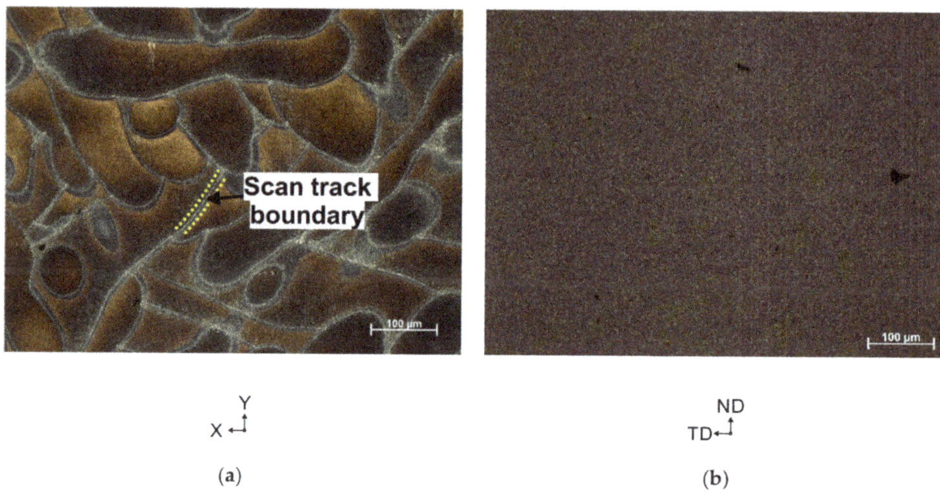

Figure 2. LM microstructures of the AlSi10Mg alloy (**a**) in the LPBF condition, (**b**) after the KoBo extrusion process (note that, in (**a**), the laser scan track boundaries are visible as the brighter areas).

Figure 2b shows the microstructure of the same AlSi10Mg alloy after undergoing the KoBo extrusion process. Here, a notable transformation is observed, as the unique laser scan tracks, which were prominent in the LPBF state, have disappeared.

A scanning electron micrograph (SEM) shows the microstructure of an AlSi10Mg in LPBF condition, in Figure 3a. As can be seen, the microstructure is characterised by a continuous cellular network of Si phases surrounding the primary Al phase. On closer examination, the microstructure exhibits heterogeneity, characterised by three distinct zones. The first corresponds to the heat-affected zone (HAZ), where the thermal impacts (increase in diffusion rate) during the LPBF process have resulted in partial rupture of the Si phase. The second zone corresponds to the coarse area of the melt pool (MP coarse), where the cellular Si network exhibits the largest size (average measured cell size of ~1.4 µm). The MP coarse area is the region that underwent an extended cooling period, which facilitates the growth of larger cells or dendrites of the Si phase (coarser structure forms where the melt pool experiences the lowest temperature gradient and the highest solidification rate) [37]. The third identified region corresponds to the MP fine area, where the cellular structures exhibit the smallest size (measured average cell size of ~0.4 µm) compared to those of the other regions. This suggests the fastest cooling rate or shortest solidification time during the LPBF process [38].

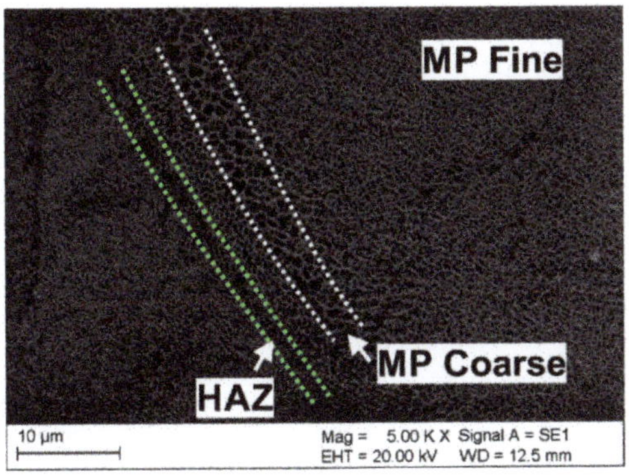

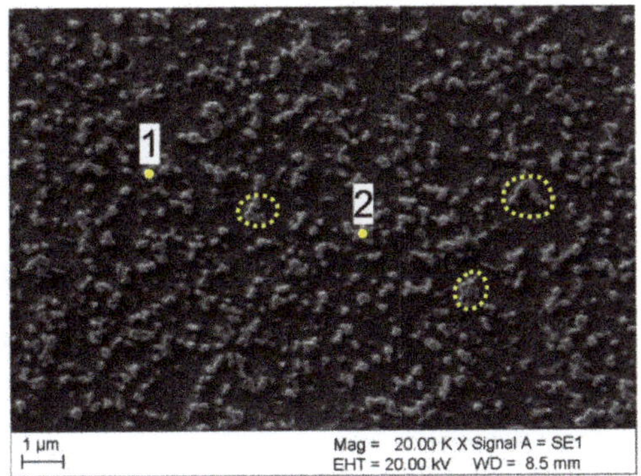

(a) (b)

Figure 3. SEM microstructures of the AlSi10Mg alloy (**a**) in the LPBF condition, (**b**) after KoBo extrusion process (note that (**a**) shows the microstructure on the "build x–y plane", whereas (**b**) shows the microstructure on the "extrusion ND–TD plane").

Figure 3b shows a detailed view of the microstructure of the KoBo sample. This SEM image reveals significant changes in the microstructure compared to the LPBF condition. It shows fine Si particles, indicating that the cellular network of the Si phase observed in the LPBF condition has been broken and transformed into discrete particles. In addition, agglomerates and clusters of silicon (Si) particles can be seen in certain areas of the SEM image (see yellow-dotted outlines) as evidenced by EDS analysis in Figure 4. The presence of these agglomerates suggests that a phenomenon known as Ostwald ripening [39] may have occurred during post-processing by KoBo extrusion.

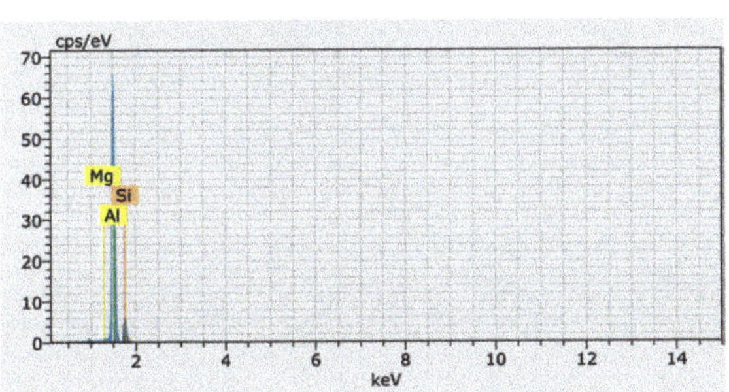

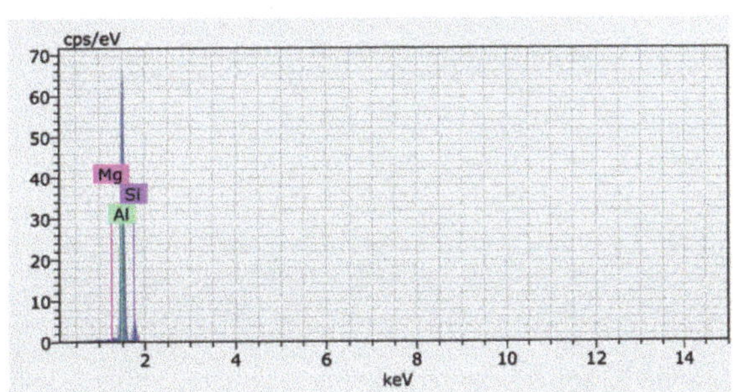

(a) (b)

Figure 4. Results of SEM EDS pointwise chemical composition analysis. (**a**) EDS spectrum corresponding to the point (1) in Figure 3b. (**b**) EDS spectrum corresponding to point (2) in Figure 3b.

The size of the individual Si particles observed in the SEM image ranges between 70 and 400 nanometres, indicating a significant reduction in size and implying a high degree of refinement accomplished during the KoBo extrusion process. The uniform distribution of Si particles in the alloy matrix indicates good mixing and dispersion during KoBo extrusion, resulting in a homogeneous microstructure. This uniform distribution can be beneficial for improving the mechanical properties and overall performance of the alloy. Remarkably, the statistical analysis performed using ImageJ software (NIG, United States) reveals that the volume fraction of the second-phase particles is approximately

20.1 ± 1%, which interestingly aligns with the volume fraction of the cellular eutectic network observed in the LPBF condition (approx. 20%).

The EBSD inverse pole figure (IPF-Z) map displays different grains with varying orientations, represented by distinct colour maps, as indicated by the unit triangle in Figure 5a. The black lines signify high-angle grain boundaries (HAGBs) with misorientations exceeding 15 degrees, while the red lines represent low-angle grain boundaries (LAGBs) with misorientations ranging from 2 to 15 degrees. The LPBF sample comprises fine equiaxed grains exhibiting relatively random orientations, with an average grain size of 6.3 μm (GTA = 5°), Figure 6a. The EBSD analysis reveals the presence of 22.8% LAGBs and 77.2% HAGBs, Table 2. The occurrence of LAGBs can be attributed to the elevated dislocation density resulting from the rapid solidification rates associated with the selective laser melting process.

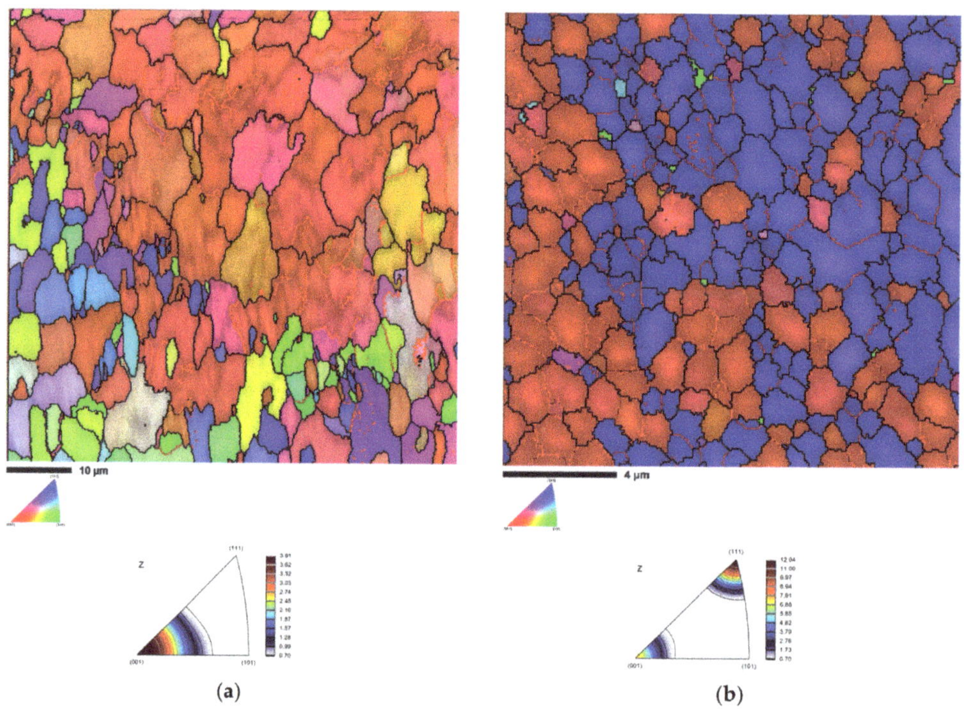

Figure 5. Results of the EBSD analysis. IPF-Z images of AlSi10Mg alloy (**a**) in the LPBF condition, (**b**) after KoBo extrusion process (red and blue colours represent the <001> and <111> fibres, respectively).

Table 2. Averaged results of microstructural analysis conducted with SEM microscope (on the basis of three separate scans).

Condition	LAGBs	HAGBs	Grain Size, μm	Recrystallised Fraction, % (Grains with GOS < 2.5°)	Geometrically Necessary Dislocation Density, m^{-2}
LPBF	22.8 ± 1.7	77.2 ± 1.2	6.3 ± 0.3	92.6 ± 2.1	1.79×10^{14}
KoBo	9.2 ± 1.3	90.8 ± 0.8	1.1 ± 0.2	93.5 ± 1.5	6.20×10^{14}

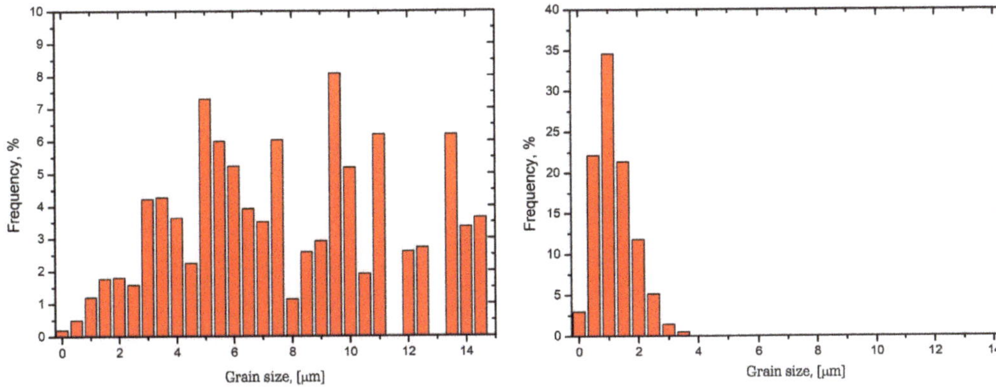

(**a**) (**b**)

Figure 6. Histograms of grain size distribution of the AlSi10Mg alloy (**a**) in the LPBF condition, (**b**) after KoBo extrusion process.

Figure 5b shows the IPF-Z map of the KoBo processed sample. The image reveals a significant occurrence of the grain refinement process following KoBo extrusion, attributed to the dynamic recrystallisation processes. The grains are significantly smaller, with an average size of approximately 1.1 μm, Figure 6b. The reduction in grain size is evident across the entire map, indicating a consistent refinement of the microstructure throughout the sample.

Furthermore, it is evident that a majority of the grains exhibit orientations aligned with the $\langle 111 \rangle_{Al}$ or $\langle 001 \rangle_{Al}$ axes, which is a characteristic feature observed in extruded alloys [40]. Most of the high-angle grain boundaries observed in the IPF map are associated with the boundaries between grains oriented along the $\langle 111 \rangle_{Al}$ or $\langle 001 \rangle_{Al}$ axes, exhibiting misorientations of approximately ~55 degrees. On the other hand, grain boundaries with low-angle misorientations are primarily linked to grains oriented along the same axis. Based on the statistical data presented in Table 2, it is observed that Low-Angle Grain Boundaries (LAGBs) account for 9.2% of the GBs, while High-Angle Grain Boundaries (HAGBs) make up the remaining 90.8% of GBs.

The GOS maps of the LPBF and KoBo samples, represented using a colour gradient, are illustrated in Figure 7a,b correspondingly. In these maps, the grains appearing in dark grey denote the smallest GOS values, indicating that they are dislocation-free and are classified as recrystallised grains. The EBSD measurements reveal that the recrystallised fractions account for 92.6% and 93.5% in the LPBF and KoBo samples, respectively.

The grain-orientation-spread (GOS) maps, showcasing the CSL boundaries highlighted with rainbow colour coding, offer valuable insights into the distribution of CSL boundaries within the sample's microstructures. Remarkably, in the LPBF state, the occurrence of low-Σ CSL boundaries is minimal, as depicted in Figure 8a. Furthermore, the Σ CSL boundaries are predominantly distributed in proximity to grains exhibiting relatively low GOS values.

After KoBo extrusion (as depicted in Figure 8b), the results indicate a notable increase in the fraction of low-Σ CSL boundaries. Specifically, the proportion of Σ3 boundaries has risen from approximately 0.6% to around 5%, while the fraction of Σ11 boundaries has increased from 1.3% to 6.3%. Furthermore, the fraction of Σ25b (nucleus of primary recrystallisation) [41] has significantly increased from 0.76% to about 7.77%. It is worth mentioning that the distribution of Σ CSL boundaries in the KoBo extruded state is similar to that observed in the LPBF state, where these boundaries are predominantly situated close to grains with low grain-orientation-spread values.

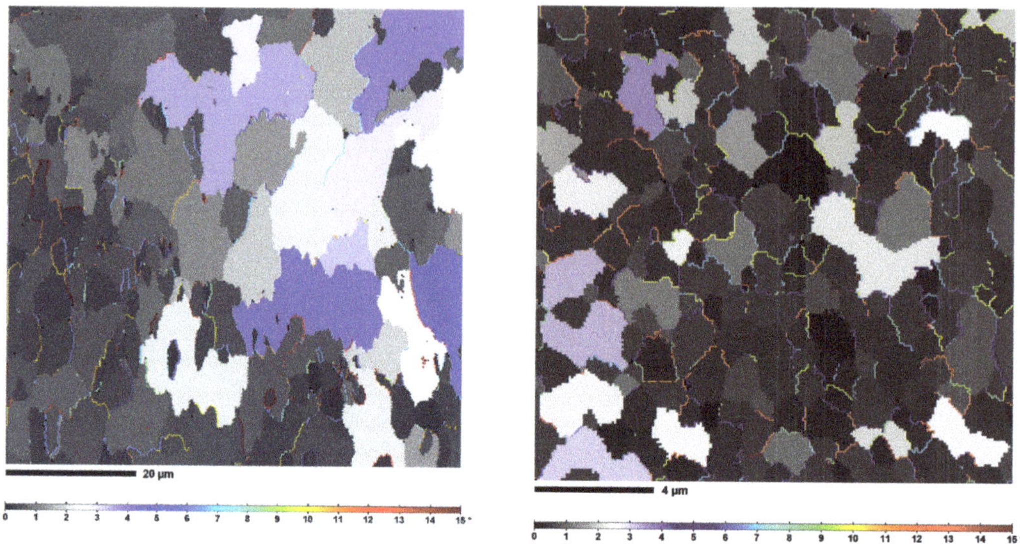

(**a**) (**b**)

Figure 7. The GOS maps of AlSi10Mg alloy (**a**) in the LPBF condition, (**b**) after KoBo extrusion process (note that in the GOS maps the CSL boundaries are shown with different colours).

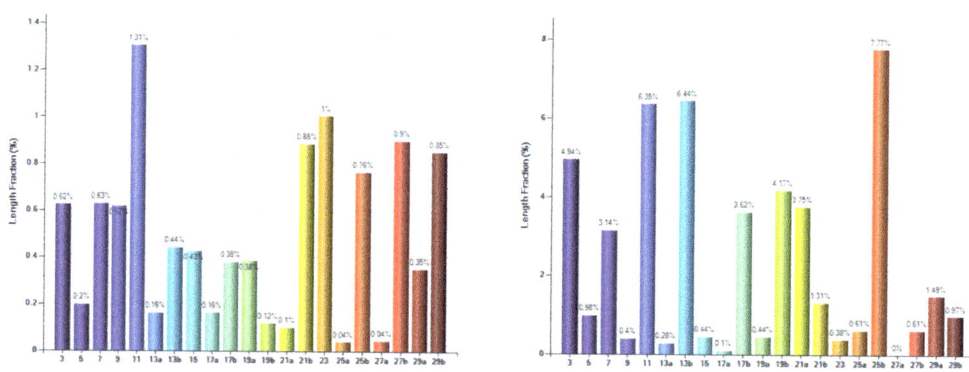

(**a**) (**b**)

Figure 8. The distribution of boundary types with respect to Σ of AlSi10Mg alloy (**a**) in the LPBF condition, (**b**) after KoBo extrusion process.

In Figure 9, the transmission Kikuchi diffraction (TKD) map provides a detailed visualisation of the grain orientation spread and the fraction of Coincidence Site Lattice (CSL) grain boundaries. This experimental result is crucial for a comprehensive characterisation of the microstructure in the KoBo sample. It is important to note that the fraction of Σ17b boundaries may be overestimated due to their detection mainly in areas with small pixel clusters (artefacts). Nevertheless, the results confirm the presence of Σ CSL boundaries in the analysed sample. Statistical analysis highlights that the highest fractions of CSL boundaries are observed for Σ3, Σ7, Σ11, and Σ13b, indicating their prevalence within the microstructure, Figure 10. Moreover, the presence of higher-order Σ9 twin boundaries indicate the occurrence of GBE mechanisms [42].

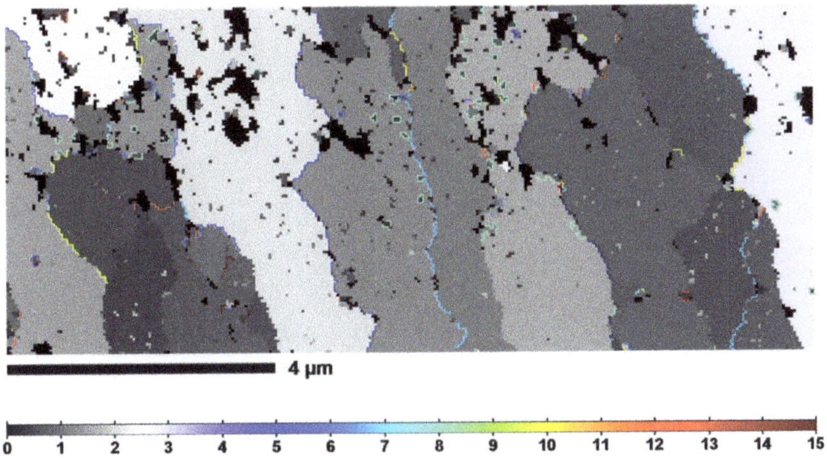

Figure 9. The TKD map of AlSi10Mg alloy showing GOS and CSL boundaries in KoBo sample.

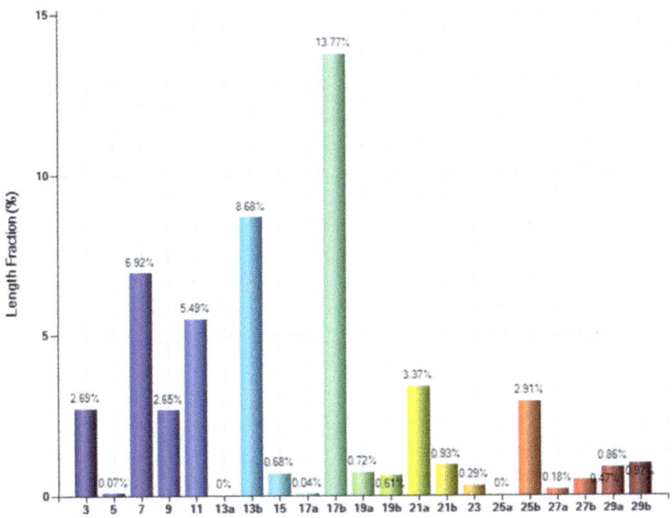

Figure 10. The distribution of boundary types with respect to Σ of AlSi10Mg alloy after KoBo extrusion process (note that this image presents the result of CSL boundary analysis received from TKD analysis).

Figure 11a shows a bright-field (BF) transmission electron micrograph of the AlSi10Mg alloy in the LPBF condition, revealing distinct microstructural features. The image shows columnar cells, elongated grain structures formed during additive manufacturing, indicating the directional solidification pattern during 3D printing. A notable aspect of the bright-field image TEM is the brightness differences in the microstructure. The columnar cell boundaries appear much brighter due to compositional differences, as they are rich in the Si phase.

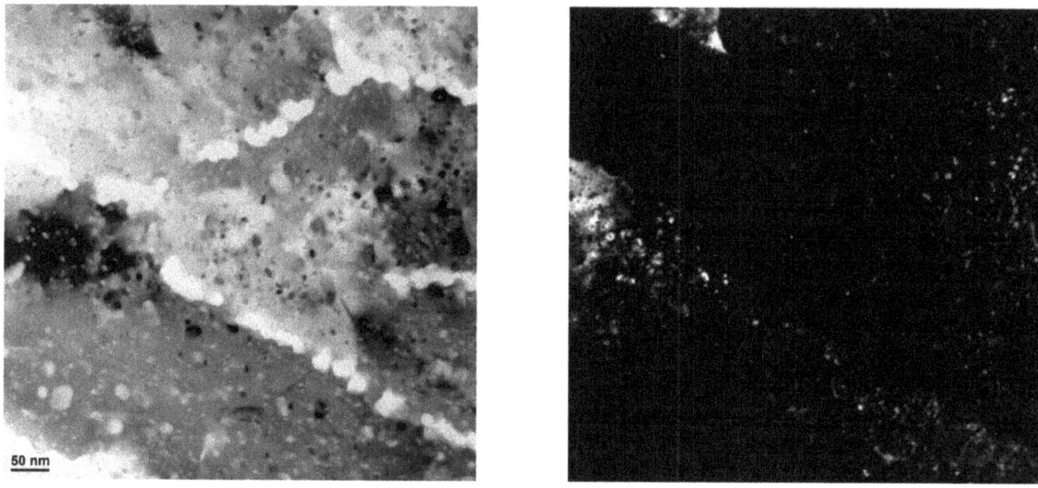

(**a**) (**b**)

Figure 11. Microstructure of AlSi10Mg alloy in LPBF condition. (**a**) Bright-field TEM image. (**b**) Dark-field TEM image.

In contrast, the dark-field TEM image emphasises the presence of dislocations in the microstructure, a common feature of additively produced Al–Si alloys, Figure 11b. The rapid solidification and thermal gradients in 3D printing can lead to a higher dislocation density compared to conventional processes.

Figure 12 shows the BF TEM micrograph of the AlSi10Mg alloy sample after the KoBo extrusion process. In this image, multiple grain boundaries are clearly visible, demarcating individual grains within the microstructure. Additionally, one particular grain stands out, exhibiting a size of approximately 300–400 nm. Importantly, the grain size measurement observed in the bright-field TEM image is comparable to the grain size results obtained through electron backscatter diffraction (EBSD) analysis, Figure 4.

The high-resolution transmission electron microscopy (HRTEM) image, Figure 12b, provides a closer look at the grain boundary highlighted in the bright-field TEM image (see black-dotted square). The boundary appears to be mostly straight but with the incorporation of some facets, as indicated by the black arrows. This HRTEM image reveals the atomic structure of the asymmetric twin boundary with a $\Sigma 3\ (221)_1/(001)_2\ \Phi = 54.56^\circ$ orientation relationship. Note that there are also several SFs that dissociated from the GB into grain at the areas near the facet junctions, which might serve to accommodate the misorientation deviation of this boundary from that of the ideal (002)/(111) ATGB (i.e., 54.56° versus 54.74°) [43].

The presence of a low-Σ value Coincidence Site Lattice (CSL) boundary indicates a low grain boundary energy, representing high fracture and corrosion resistance [44]. It should be pointed out that the presence of a CSL grain boundary observed through HRTEM aligns with the Electron Backscatter Diffraction (EBSD) data.

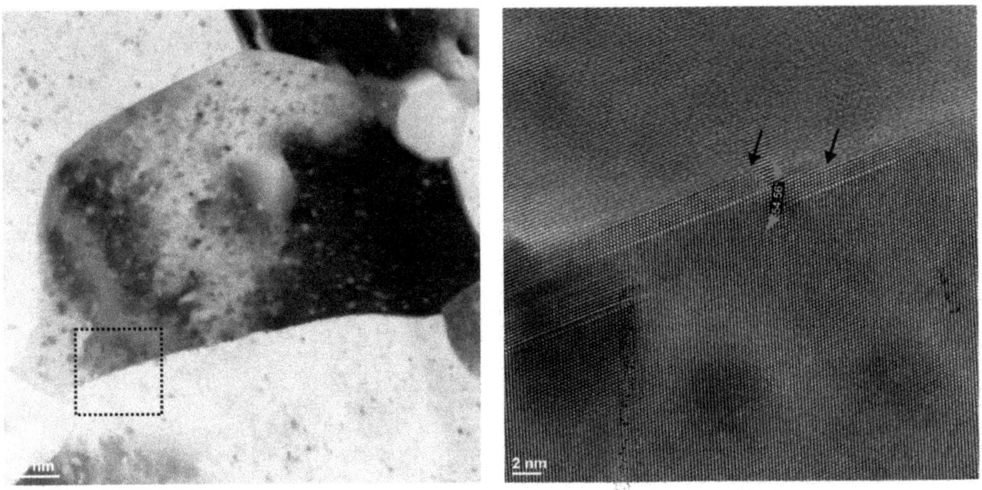

(a) (b)

Figure 12. Microstructure of AlSi10Mg alloy subjected to KoBo extrusion. (**a**) Bright-field TEM image. (**b**) HRTEM image of the boundary area highlighted in (**a**).

4. Discussion

4.1. CSL Boundary Formation

In this study, we investigated the formation of CSL (Coincidence Site Lattice) boundaries in a sample subjected to the KoBo extrusion process. The results revealed a significant increase in the fraction of low-Σ CSL boundaries, particularly Σ3, Σ7 and Σ11 boundaries, after the KoBo extrusion compared to the LPBF state. Additionally, the fraction of Σ25b (nucleus of primary recrystallisation) exhibited a remarkable increase. These observations prompt an exploration of the possible mechanisms underlying the formation of CSL boundaries in the KoBo-extruded sample.

One potential explanation lies in the deformation and recrystallisation processes during KoBo extrusion, where the plastic deformation leads to the generation of dislocations and the rearrangement of crystal grains. The subsequent recrystallisation may favour the nucleation of grains with low orientation spread values, which could coincide with the observed proximity of Σ CSL boundaries to these grains. In this regard, the formation of CSL boundaries can be attributed to the atomic interactions that occur as a result of a dynamic recrystallisation-driven grain boundary migration process [45]. Alternatively, it is also plausible that the rise in the fraction of low-Σ Coincidence Site Lattice boundaries is a consequence of interactions between the pre-existing CSL boundaries during recrystallisation [46]. These interactions may have facilitated the formation of new CSL boundaries, resulting in a higher prevalence of low-Σ CSL boundaries.

Furthermore, in the context of severe plastic deformation, a phenomenon known as grain boundary (GB) relaxation may occur when the grain size decreases below a critical threshold. The mechanism underlying this GB relaxation in nanostructured materials is believed to be triggered by the emission of partial dislocations and stacking faults from the grain boundaries [47]. It has also been shown that Si has the ability to lower the stacking fault energy [48], which might reduce the critical shear stress required to nucleate partial dislocations. Moreover, the solute atoms increase the localised stress needed for dislocation motion, thus reducing annihilation and rearrangement of dislocation.

Recent experimental evidence also points to the significance of strain-induced boundary migration (SIBM) in enhancing the frequency and distribution of CSL boundaries [49–51]. While the migration of grain boundaries during primary recrystallisation involves the formation and growth of nuclei with different orientations, SIBM leads to the creation of stress-free regions with the same orientation as the parent grain [52]. These stress-free orientation regions can be analysed using the density of geometrically necessary dislocations (GNDs).

Upon analysis, it is observed that in the KoBo processed sample, CSL boundaries are predominantly distributed in proximity to regions with a higher density of GNDs (depicted as red "hot spots" in Figure 13). Consequently, this specific mechanism can be considered negligible, as twinning occurs most frequently at the points of highest GND accumulation.

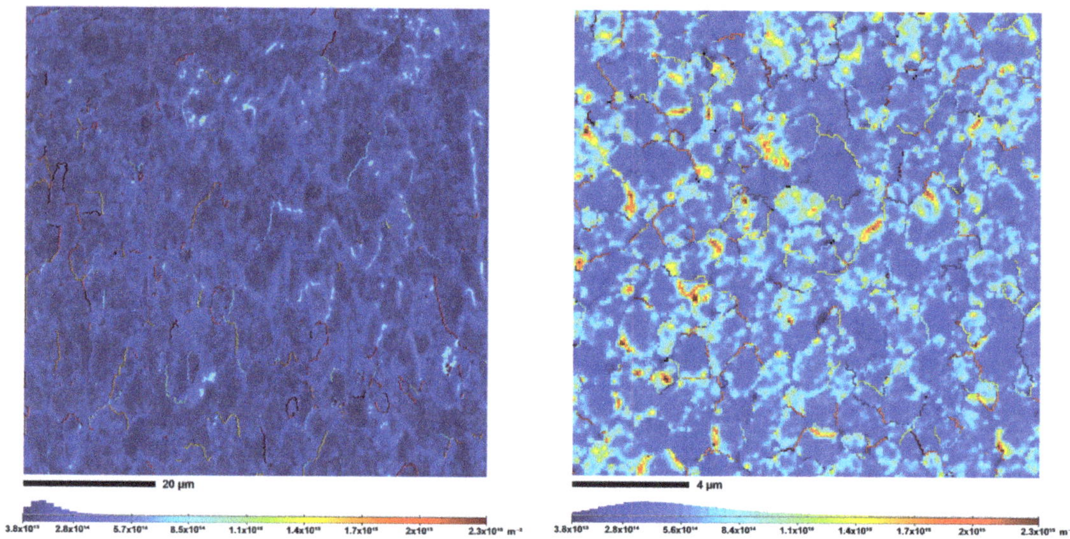

(a) (b)

Figure 13. GND distribution maps (**a**) in the LPBF condition, (**b**) after KoBo extrusion process.

4.2. Texture Analysis

As shown in Figure 5, the grains of the AlSi10Mg alloy show preferred crystallographic orientations. In the case of the LPBF state, the majority of grains have a strong texture along (100) direction, Figure 5a. The (100), (110), (111) and (112) pole figures, Figure 14a confirm that the ⟨100⟩ cube texture is the predominant in the LPBF condition, which is typical for LPBF Al–Si alloys [53,54].

Previous study [55] demonstrated that severe cyclic torsion in KoBo extrusion process can induce the development of a strong texture characterised by the presence of intense poles corresponding to specific orientations of crystallites. As can be seen in Figure 14b, the texture of the AlSi10Mg alloy subjected to KoBo extrusion is consistent with the models of the authors of the manuscript cited above. A great majority of the crystallites display (111) and (100) orientations, corresponding to ⟨111⟩ and ⟨001⟩ fibre-type texture components. Moreover, it is evident that certain crystallites exhibit a weak cube texture, which is indicative of the recrystallisation process [56].

The recrystallisation texture displays an axial pattern, which can be attributed to the formation of recrystallisation growth nuclei with orientations near ⟨100⟩ and ⟨111⟩ in the deformed grains exhibiting corresponding orientations. These nuclei grow within the surrounding deformed matrix when their misorientation with the neighbouring area aligns at an angle close to 52° around one of the crystallographic axes, specifically ⟨331⟩ [41]. As a result, the misorientation closely approximates the CSL (Coincidence Site Lattice) misorientation Σ25b (see Figure 8b).

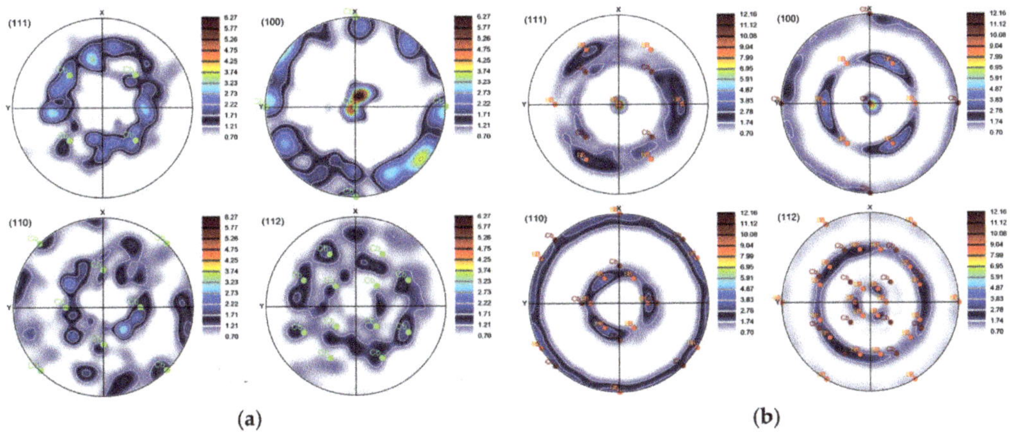

Figure 14. Pole figures of the AlSi10Mg alloy (**a**) in LPBF condition, (**b**) after KoBo extrusion process.

5. Conclusions

1. The KoBo extrusion process substantially refined the heterogeneous microstructure of an LPBF AlSi10Mg aluminium alloy. Also, the cellular microstructure diminished and was replaced by the fine Si precipitates distributed uniformly within the aluminium matrix.
2. The microstructure of the LPBF material underwent significant refinement as a result of the KoBo extrusion process, reducing the grain size to the sub-micrometric level. The study revealed that the KoBo sample displayed an average grain size of 1.1 μm. This refined microstructure is expected to have a substantial impact on the material's properties.
3. The KoBo extrusion process led to a severe increment of the high-angle grain boundary population, indicating the dynamic recrystallisation process.
4. The analyses of the IPF images revealed grains oriented along the $\langle 111 \rangle$ and $\langle 001 \rangle$ axes and a high number of recrystallised grains with a near $\{111\}$ $\langle 112 \rangle$ orientation.
5. High-strain and high-rate deformation post-processing of the LPBF AlSi10Mg alloy resulted in a grain-boundary-engineered microstructure, as evidenced by increased length fractions of the $\Sigma 3$, $\Sigma 7$ and $\Sigma 11$ boundaries. It was found that the fraction of CSL boundaries increases after the KoBo extrusion process.

Author Contributions: P.S.: conceptualization, methodology, validation, formal analysis, resources, data curation, writing—original draft, visualization, supervision, project administration, founding acquisition; K.M.: data curation, formal analysis. P.S. contributed 85% of the research work, primarily by conducting the KoBo experiments, microstructural analysis, data collection, data analysis, project administration and writing of the manuscript; K.M. contributed 15% of research work, primarily by conducting the microstructural analysis. All authors have read and agreed to the published version of the manuscript.

Funding: The research was funded by the National Science Centre, Poland, based on the decision number 2021/43/D/ST8/01946.

Institutional Review Board Statement: Not applicable.

Informed Consent Statement: Not applicable.

Data Availability Statement: Data available on request.

Conflicts of Interest: The authors declare no conflict of interest.

References

1. Jiang, Y.; Moubayidin, L. Floral symmetry: The geometry of plant reproduction. *Emerg. Top. Life Sci.* **2022**, *6*, 259–269. [CrossRef] [PubMed]
2. Wu, X.; Knuth, M.J.; Hall, C.R.; Palma, M.A. Increasing Profit Margins by Substituting Species in Floral Arrangements. *Horttechnology* **2021**, *31*, 19–26. [CrossRef]
3. Zhao, P.; Zhang, K.; Zhao, C.; Qi, L.; Deng, Z. In-plane wave propagation analysis for waveguide design of hexagonal lattice with Koch snowflake. *Int. J. Mech. Sci.* **2021**, *209*, 106724. [CrossRef]
4. Snyder, M.; Kara, M. A Modeling Activity: Analyzing Snowflakes Using Geometry. *Ohio J. Sch. Math.* **2022**, *92*, 1–7.
5. Pignataro, T.; Lourenço, G.M.; Beirão, M.; Cornelissen, T. Wings are not perfect: Increased wing asymmetry in a tropical butterfly as a response to forest fragmentation. *Sci. Nat.* **2023**, *110*, 28. [CrossRef] [PubMed]
6. Shamir, L. Using Machine Learning to Profile Asymmetry between Spiral Galaxies with Opposite Spin Directions. *Symmetry* **2022**, *14*, 934. [CrossRef]
7. Lovell, T.C.; Colwell, C.E.; Zakharov, L.N.; Jasti, R. Symmetry breaking and the turn-on fluorescence of small, highly strained carbon nanohoops. *Chem. Sci.* **2019**, *10*, 3786–3790. [CrossRef]
8. Votano, J.; Parham, M.; Hall, L. *Handbook of Aluminum Volume 2 Alloy Production and Material Manufacturing*; CRC Press: Boca Raton, FL, USA, 2004; pp. 1–731.
9. Gargeya, B.S.K.; Babu, P.N.; Pal, S. Constant twist rate response of symmetric and asymmetric Σ5 aluminium tilt grain boundaries: Molecular dynamics study of deformation processes. *J. Mater. Sci.* **2021**, *56*, 8544–8562. [CrossRef]
10. Tschopp, M.A.; Coleman, S.P.; McDowell, D.L. Symmetric and asymmetric tilt grain boundary structure and energy in Cu and Al (and transferability to other fcc metals). *Integr. Mater. Manuf. Innov.* **2015**, *4*, 176–189. [CrossRef]
11. Mishin, Y.; Asta, M.; Li, J. Atomistic modeling of interfaces and their impact on microstructure and properties. *Acta Mater.* **2010**, *58*, 1117–1151. [CrossRef]
12. Zhang, C.; Lin, L.; Chen, R.; Zhang, L.; Shao, Z. Grain Boundary Engineering and Its Effect on Intergranular Corrosion Resistance of a Ni-Cr-Mo Based C276 Superalloy. *Crystals* **2022**, *12*, 1625. [CrossRef]
13. Dolzhenko, P.; Tikhonova, M.; Odnobokova, M.; Kaibyshev, R.; Belyakov, A. On Grain Boundary Engineering for a 316L Austenitic Stainless Steel. *Metals* **2022**, *12*, 2185. [CrossRef]
14. Laleh, M.; Hughes, A.E.; Tan, M.Y.; Rohrer, G.S.; Primig, S.; Haghdadi, N. Grain boundary character distribution in an additively manufactured austenitic stainless steel. *Scr. Mater.* **2021**, *192*, 115–119. [CrossRef]
15. Randle, V. Relationship between coincidence site lattice, boundary plane indices, and boundary energy in nickel. *Mater. Sci. Technol.* **1999**, *15*, 246–252. [CrossRef]
16. Jiang, X.; Yang, C.; Zhang, W.; Wang, X. Surface grain boundary engineering in 304 stainless steel by means of mechanical grinding treatment-induced gradient plastic strain and annealing. *J. Mater. Sci.* **2022**, *57*, 21798–21812. [CrossRef]
17. Guan, X.J.; Shi, F.; Jia, Z.P.; Li, X.W. Grain boundary engineering of AL6XN super-austenitic stainless steel: Distinctive effects of planar-slip dislocations and deformation twins. *Mater. Charact.* **2020**, *170*, 110689. [CrossRef]
18. Chen, X.-H.; Wang, F.; Zhang, F. Grain boundary engineering process for nano reinforced aluminum matrix composites. *J. Alloys Compd.* **2023**, *939*, 168834. [CrossRef]
19. Shi, Z.; He, R.; Chen, Y.; Yan, H.; Song, H.; Luo, C.; Nie, Q.; Hu, Z. Microstructural evolution and strengthening mechanisms of a novel Al–11Si–3Cu alloy microalloyed with minor contents of Sr and Sc. *Mater. Sci. Eng. A* **2022**, *853*, 143738. [CrossRef]
20. Richter, N.A.; Zhang, Y.F.; Xie, D.Y.; Su, R.; Li, Q.; Xue, S.; Niu, T.; Wang, J.; Wang, H.; Zhang, X. Microstructural evolution of nanotwinned Al-Zr alloy with significant 9R phase. *Mater. Res. Lett.* **2021**, *9*, 91–98. [CrossRef]
21. Liu, M.; Wang, P.; Lu, G.; Huang, C.-Y.; You, Z.; Wang, C.-H.; Yen, H.-W. Deformation-activated recrystallization twin: New twinning path in pure aluminum enabled by cryogenic and rapid compression. *iScience* **2022**, *25*, 104248. [CrossRef]
22. Guo, B.; Song, M.; Zhang, X.; Liu, Y.; Cen, X.; Chen, B.; Li, W. Exploiting the synergic strengthening effects of stacking faults in carbon nanotubes reinforced aluminum matrix composites for enhanced mechanical properties. *Compos. Part B Eng.* **2021**, *211*, 108646. [CrossRef]
23. Gong, D.; Cao, Y.; Deng, X.; Jiang, L. Higher proportional limit of SiC/Al composites with nano-scaled stacking faults. *Compos. Commun.* **2022**, *32*, 101188. [CrossRef]
24. Xue, S.; Fan, Z.; Lawal, O.B.; Thevamaran, R.; Li, Q.; Liu, Y.; Yu, K.Y.; Wang, J.; Thomas, E.L.; Wang, H.; et al. High-velocity projectile impact induced 9R phase in ultrafine-grained aluminium. *Nat. Commun.* **2017**, *8*, 1653. [CrossRef] [PubMed]
25. Xiong, Q.; Li, Z.; Huang, X.; Shimada, T.; Kitamura, T. Thermomechanical conversion in high-rate plastic deformation of nanotwinned polycrystalline copper. *J. Therm. Stress.* **2022**, *45*, 65–80. [CrossRef]
26. Qian, L.; Cui, Z.; Sun, C.; Geng, S.; Sun, Z. Investigation of deformation compatibility and power consumption during KOBO extrusion of bimetallic composite tube. *Int. J. Adv. Manuf. Technol.* **2022**, *118*, 3477–3486. [CrossRef]
27. Valiev, R.Z.; Langdon, T.G. Principles of equal-channel angular pressing as a processing tool for grain refinement. *Prog. Mater. Sci.* **2006**, *51*, 881–981. [CrossRef]
28. Snopiński, P.; Matus, K.; Tatiček, F.; Rusz, S. Overcoming the strength-ductility trade-off in additively manufactured AlSi10Mg alloy by ECAP processing. *J. Alloys Compd.* **2022**, *918*, 165817. [CrossRef]

29. Snopiński, P.; Król, M.; Pagáč, M.; Petrů, J.; Hajnyš, J.; Mikuszewski, T.; Tański, T. Effects of equal channel angular pressing and heat treatments on the microstructures and mechanical properties of selective laser melted and cast AlSi10Mg alloys. *Arch. Civ. Mech. Eng.* **2021**, *21*, 92. [CrossRef]
30. Zhilyaev, A.P.; Langdon, T.G. Using high-pressure torsion for metal processing: Fundamentals and applications. *Prog. Mater. Sci.* **2008**, *53*, 893–979. [CrossRef]
31. Tsuji, N.; Saito, Y.; Lee, S.-H.; Minamino, Y. ARB (Accumulative Roll-Bonding) and other new Techniques to Produce Bulk Ultrafine Grained Materials. *Adv. Eng. Mater.* **2003**, *5*, 338–344. [CrossRef]
32. Korbel, A.; Błaż, L.; Bochniak, W.; Pawlyta, M.; Ostachowski, P.; Łagoda, M. Nano-Dimensional Elements in the Structure of Zinc Subjected to KOBO Extrusion. *Met. Microstruct. Anal.* **2023**, *12*, 427–432. [CrossRef]
33. Dai, C.; Saidi, P.; Yao, Z.; Béland, L.K.; Daymond, M.R. Deformation-free nanotwin formation in zirconium and titanium. *Mater. Lett.* **2019**, *247*, 111–114. [CrossRef]
34. Bai, F.M.; Ye, X.; Zhang, H.Y.; Zhou, H.W.; Song, M.; Sun, Y.X.; He, Y.Z. A significant increase in the hardness of nanotwinned titanium alloys prepared via the martensitic phase transformation. *Mater. Lett.* **2019**, *255*, 126507. [CrossRef]
35. Bochniak, W.; Marszowski, K.; Korbel, A. Theoretical and practical aspects of the production of thin-walled tubes by the KOBO method. *J. Mater. Process. Technol.* **2005**, *169*, 44–53. [CrossRef]
36. Beausir, B.; Fundenberger, J.J. *Analysis Tools for Electron and X-ray Diffraction, ATEX–Software*; Université de Lorraine-Metz: Nancy, France, 2007.
37. Rafieazad, M.; Mohammadi, M.; Nasiri, A.M. On microstructure and early stage corrosion performance of heat treated direct metal laser sintered AlSi10Mg. *Addit. Manuf.* **2019**, *28*, 107–119. [CrossRef]
38. Thijs, L.; Kempen, K.; Kruth, J.P.; Van Humbeeck, J. Fine-structured aluminium products with controllable texture by selective laser melting of pre-alloyed AlSi10Mg powder. *Acta Mater.* **2013**, *61*, 1809–1819. [CrossRef]
39. Ding, C.; Hao, H.; Lu, Z.; Yu, C.; Wu, X.; Yu, P.; Ye, S. Fabrication of hypereutectic Al–Si alloy with improved mechanical and thermal properties by hot extrusion. *Mater. Charact.* **2023**, *202*, 113026. [CrossRef]
40. Nittala, A.; Smith, J.; Gwalani, B.; Silverstein, J.; Kraft, F.F.; Kappagantula, K. Simultaneously improved electrical and mechanical performance of hot-extruded bulk scale aluminum-graphene wires. *Mater. Sci. Eng. B* **2023**, *293*, 116452. [CrossRef]
41. Danilov, S.V.; Reznik, P.L.; Zorina, M.A.; Lobanov, M.L. Effect of special boundaries on recrystallization texture of FCC metals with high packing defect energy. *AIP Conf. Proc.* **2019**, *2174*, 20207. [CrossRef]
42. Jones, R.; Randle, V. Sensitisation behaviour of grain boundary engineered austenitic stainless steel. *Mater. Sci. Eng. A* **2010**, *527*, 4275–4280. [CrossRef]
43. Fang, Z.; Xiao, J.; Tan, S.; Deng, C.; Wang, G.; Mao, S.X. Atomic-scale observation of dynamic grain boundary structural transformation during shear-mediated migration. *Sci. Adv.* **2023**, *8*, eabn3785. [CrossRef] [PubMed]
44. Oda, S.; Tanaka, S.-I. Grain boundaries with high Σ value and strain in grain matrix induce crack initiation in extruded 6000 series aluminium alloys. *Mater. Sci. Eng. A* **2022**, *834*, 142630. [CrossRef]
45. Cao, Y.; Di, H.; Zhang, J.; Zhang, J.; Ma, T.; Misra, R.D.K. An electron backscattered diffraction study on the dynamic recrystallization behavior of a nickel–chromium alloy (800H) during hot deformation. *Mater. Sci. Eng. A* **2013**, *585*, 71–85. [CrossRef]
46. Mandal, S.; Jayalakshmi, M.; Bhaduri, A.K.; Subramanya Sarma, V. Effect of Strain Rate on the Dynamic Recrystallization Behavior in a Nitrogen-Enhanced 316L(N). *Met. Mater. Trans. A* **2014**, *45*, 5645–5656. [CrossRef]
47. Zhu, R.J.; Zhou, X.; Li, X.Y. Thermal stability of nanograins with grain boundary relaxation in microalloyed Cu-Sb and Cu-Fe. *J. Mater. Sci. Technol.* **2023**, *155*, 66–71. [CrossRef]
48. Muzyk, M.; Pakiela, Z.; Kurzydlowski, K.J. Ab initio calculations of the generalized stacking fault energy in aluminium alloys. *Scr. Mater.* **2011**, *64*, 916–918. [CrossRef]
49. Garcia-Chao, P.; Eipe, J.J.; Krugla, M.; Bos, C.; Sietsma, J.; Kranendonk, W.; Offerman, S.E. Nucleation Sites in the Static Recrystallization of a Hot-Deformed Ni-30 Pct Fe Austenite Model Alloy. *Met. Mater. Trans. A* **2023**, *54*, 2160–2177. [CrossRef]
50. Detrois, M.; McCarley, J.; Antonov, S.; Helmink, R.C.; Goetz, R.L.; Tin, S. Comparative study of high-temperature grain boundary engineering of two powder-processed low stacking-fault energy Ni-base superalloys. *Mater. High Temp.* **2016**, *33*, 310–317. [CrossRef]
51. Sarkari Khorrami, M.; Saito, N. On the formation of large grain structure after friction stir processing of ultrafine-grained aluminium alloy. *Philos. Mag.* **2023**, *103*, 733–748. [CrossRef]
52. Min, K.M.; Jeong, W.; Hong, S.H.; Lee, C.A.; Cha, P.-R.; Han, H.N.; Lee, M.-G. Integrated crystal plasticity and phase field model for prediction of recrystallization texture and anisotropic mechanical properties of cold-rolled ultra-low carbon steels. *Int. J. Plast.* **2020**, *127*, 102644. [CrossRef]
53. Zhao, L.; Song, L.; Santos Macías, J.G.; Zhu, Y.; Huang, M.; Simar, A.; Li, Z. Review on the correlation between microstructure and mechanical performance for laser powder bed fusion AlSi10Mg. *Addit. Manuf.* **2022**, *56*, 102914. [CrossRef]
54. Sathishkumar, A.; Soundararajan, R.; Sivasankaran, S. Effect of Direct Aging on the Microstructure and Mechanical Behavior of AlSi10Mg Alloy: Casting Versus Selective Laser Melting. *J. Mater. Eng. Perform.* **2023**, *32*, 3215–3229. [CrossRef]

55. Kowalczyk-Gajewska, K.; Stupkiewicz, S. Modelling of Texture Evolution in Kobo Extrusion Process/Modelowanie Rozwoju Tekstury W Procesie Wyciskania Metoda Kobo. *Arch. Metall. Mater.* **2013**, *58*, 113–118. [CrossRef]
56. Lee, D.N.; Han, H.N. *Recrystallization Textures of Metals and Alloys*; Wilson, P., Ed.; IntechOpen: Rijeka, Croatia, 2013; Chapter 1.

symmetry

MDPI

Article

Investigating Cathode Electrolyte Interphase Formation in NMC 811 Primary Particles through Advanced 4D-STEM ACOM Analysis

Kevyn Gallegos-Moncayo [1,2], Justine Jean [1,3], Nicolas Folastre [1,3], Arash Jamali [1,4] and Arnaud Demortière [1,2,3,*]

1 Laboratoire de Réactivité et Chimie des Solides (LRCS), CNRS UMR 7314, 80009 Amiens, France
2 ALISTORE-European Research Institute, CNRS FR 3104, 80039 Amiens, France
3 Réseau sur le Stockage Electrochimique de L'energie (RS2E), CNRS FR 3459, 80009 Amiens, France
4 Plateforme de Microscopie Electronique, UPJV, Hub de L'energie, 15 rue Baudelocque, 80039 Amiens, France
* Correspondence: arnaud.demortiere@cnrs.fr

Abstract: This study focuses on NMC 811 ($LiNi_{0.8}Mn_{0.1}Co_{0.1}O_2$), a promising material for high-capacity batteries, and investigates the challenges associated with its use, specifically the formation of the cathode electrolyte interphase (CEI) layer due to chemical reactions. This layer is a consequence of the position of the Lowest Unoccupied Molecular Orbital (LUMO) energy level of NMC 811 that is close to the Highest Occupied Molecular Orbital (HOMO) level of liquid electrolytes, resulting in electrolyte oxidation and cathode surface alterations during charging. A stable CEI layer can mitigate further degradation by reducing the interaction between the reactive cathode material and the electrolyte. Our research analyzed the CEI layer on NMC 811 using advanced techniques, such as 4D-STEM ACOM (automated crystal orientation mapping) and STEM-EDX, focusing on the effects of different charging voltages (4.3 V and 4.5 V). The findings revealed varying degrees of degradation and the formation of a fluorine-rich layer on the secondary particles. Detailed analysis showed that the composition of this layer differed based on the voltage: only LiF at 4.5 V and a combination of lithium fluoride (LiF) and lithium hydroxide (LiOH) at 4.3 V. Despite LiF's known stability as a CEI protective layer, our observations indicate that it does not effectively prevent degradation in NMC 811. The study concluded that impurities and unwanted chemical reactions leading to suboptimal CEI formation are inevitable. Therefore, future efforts should focus on developing protective strategies for NMC 811, such as the use of specific additives or coatings.

Keywords: NMC 811; CEI layer; 4D-STEM ACOM; STEM-EDX; cathode primary particles

1. Introduction

Today, the necessity of battery materials has increased immensely due to the growing portable devices market and the energetic transition from fossil fuels to electric vehicles (EV) [1,2]. This exponential growth has pushed the scientific community to search for new materials with an acceptable cost/safety ratio in lithium-ion batteries (LIBs) to be part of this energetic change. One of the battery materials that has been studied is layered cathode materials ($LiMO_2$, M transition metal), such as NMC ($LiNiMnCoO_2$). NMC is based on an existing cathode material, $LiCoO_2$, in which Co is partially replaced by Ni and Mn, as a result obtaining $LiNi_xMn_yCo_zO_2$ ($x + y + z = 1$) [3]. In the case of NMC 811, the predominant transition metal (TM) is Ni, with the final chemical composition being $LiNi_{0.8}Mn_{0.1}Co_{0.1}O_2$ (Ni-rich cathode material). NMC 811 is a layered material with a space group R-3m (lattice parameters a/b = 2.871 Å, c = 14.20 Å [4]). Besides the energy density, the objective of this replacement is to reduce the quantity of Co because of its cost, safety, and difficulties to recycle [5]. NMC 811 presents a high discharge capacity thanks to its Ni content, with theoretical values of ~200 mAh/g [4,6].

The NMC cathode is usually used at a high voltage window (over 4.2 V) and presents capacity fading at a low cycle number due to different degradation mechanisms [4,7,8]. During the charge of batteries at high voltage, the phenomenon takes place at the interface of the liquid electrolyte and solid active material interface, leading to the CEI formation. At a high stage of Li removal, the chemical potential of the cathode material is shifted and approaches the HOMO energy of the electrolyte. This leads to its oxidation and the formation of a CEI layer, which has an impact on Li transport as well as the electronic configuration of TM sites in the cathode. The layer is composed of lithium carbonates, oxides, alkyl carbonates, as a result of solvent electrolyte oxidation parasitic reactions, and $Li_XPO_YF_Z$ oxidation products coming from the electrolyte salt. It has been shown that for Ni-rich cathode materials, it is important to take care of the LUMO/HOMO energies of the electrode and the electrolyte, as well as the nucleophilic affinity of the components [9–12].

In the study led by Iban Azcarate et al. [13], the reactivity of the LP30 electrolyte ($LiPF_6$ salt in a di-methyl carbonate/ethyl carbonate mixture), commonly utilized in LIBs, was explored. Using both simple- and double-cell configurations with glassy carbon cathodes and lithium metal anodes, the electrolyte's behavior was scrutinized through NMR (nuclear magnetic resonance) and XPS (X-ray photoelectron spectroscopy). The research revealed that at 4.2 V, ethyl carbonate (EC) was the initial electrolyte component to undergo oxidation. When the voltage increased to 4.8 V, di-methyl carbonate (DMC) also began oxidizing, generating various derivative products. These products underwent further oxidation at 5.4 V. The study also observed depositions of inorganic species, such as LiF at 4.2 V, with the predominant deposition comprising mostly organic products, forming a non-passivating layer between 4.2 and 4.8 V. Above 4.8 V, the CEI layer predominantly consisted of inorganic compounds, enhancing passivation, although it remained unstable up to 5.4 V. Notably, the researchers achieved increased passivity and stability of the cell by maintaining it at a constant voltage for several hours.

In NMC cathode materials, capacity increases as a function of the amount of Ni. However, to reach the theoretical high capacity of Ni-rich cathode materials, it is necessary to charge the battery up to a voltage above 4.2 V, leading to degradations of the common LP30 electrolyte. In the work of Noh et al. [8], a quasi-linear correlation between the Ni content increasing and a decrease in the safety and stability of the battery have been clearly demonstrated. This phenomenon can be attributed to changes in the microstructure and chemical properties that correspond to alterations in the nickel content. In cathode materials with a high nickel content, it has been observed that during cycling in the LP30 electrolyte, the instability of the CEI layer leads to a reaction between Ni^{4+} and the electrolyte at an advanced de-lithiation stage. This reaction amplifies the Li^+/Ni^{2+} cation mixing, thereby accelerating structural deterioration. Furthermore, structural decay is also exacerbated by changes in the lattice parameters, particularly along the c-axis. These changes induce strains that cause ruptures in secondary particles, which then come into contact with the electrolyte, further contributing to the material's degradation [14–17].

The CEI presence over NMC 811 particles has been reported in the literature multiple times, as well as its importance for the correct performance of LIBs. Parasitic reactions and products reduce the available Li quantity in the cell, and due to the non-ionic conductivity of these products, impede the utilization of the remaining lithium, ultimately leading to a decreased battery capacity [18–22]. One of the strategies for reducing these reactions is the formation of an ionic conductive layer, stable during cycling and resistant to mechanical deformation. Some works focused on the introduction of additives to the electrolyte for the formation of a stable CEI layer. In the work of Sen et al. [20], triallylamine (TAA) was proposed as an additive to eliminate the presence of parasite compounds, such as HF, that cause cathode damage. In this work, the batteries with the TAA electrolyte additive presented better performances in capacity retention, the CEI layer was more uniform and compact, and there was no presence of cracks on the surface of NMC 811 particles.

Another strategy found in the literature is NMC 811 coating [12,23]. Looking deeply at the work of Bishnu P. et al. [12], it was proposed to apply the electrochemical fluorination technique (ECF) to form a stable LiF layer at the surface of the particles. Pristine NMC 811 forms a non-stable CEI layer composed of inorganic compounds, such as LiOH, Li_2CO_3, and Li_2O. These compounds are electric and ionic insulators, which result in a decrease of battery capacity due to the inaccessibility to Li ions on the particles. The LiF layer is an ionic conductor as well as an electric insulator, avoiding the degradation of LP30 due to electrochemical reactions, but allowing the transportation of Li ions between the cathode and the electrolyte. The results showed that the formation of a stable CEI at low voltages increased the cycling stability of NMC 811.

TEM methodologies have been extensively employed to conduct in-depth investigations of battery materials at various scales [24–27]. Notably, the 4D-STEM ACOM technique has been instrumental in achieving precise phase characterization within batteries, both in situ and postmortem. This local analytical method offers a significant advantage in detecting compounds through electron diffraction patterns, as opposed to solely conducting elemental analysis (i.e., detecting individual species). Furthermore, 4D-STEM strikes an optimal balance between accessibility and resolution, especially when juxtaposed with other compound detection methods in battery research, such as neutron diffraction or XPS [28–32].

In the study by Ankush et al. [33], the researchers investigated LMNO (lithium–manganese–nickel–oxide) thin films in situ and postmortem, utilizing a specially adapted liquid TEM sample holder for electrochemical analysis. This investigation revealed insightful details about the electrochemical behavior of LMNO, including the characteristic oxidation peaks of nickel. Additionally, it highlighted the coexistence of amorphous and crystalline phases in LMNO and identified the formation of organic compounds resulting from electrolyte degradation. Despite the versatility of the 4D-STEM ACOM technique, it encounters limitations in liquid cell environments, primarily due to signal-to-noise reduction caused by multi-scattering effects from the liquid electrolyte's thickness. However, these challenges can be mitigated through applications such as ePattern suite software (1.1 version), as demonstrated in the work of Folastre et al. [34]. This algorithm employs registration and reconstruction methods to enhance the pattern identification and denoising of diffraction signals. Such advancements are pivotal in improving the image quality and signal-to-noise ratio, thereby enabling more reliable and accurate pattern analysis in TEM studies.

The objective of this research is to investigate the intrinsic characteristics and the genesis of the CEI in NMC 811coin cells through postmortem examination. This investigation employs an integrated approach utilizing SEM-EDX for the analysis of secondary particles, and STEM-EDX in conjunction with 4D-STEM ACOM for the examination of primary particles. This analysis is further correlated with the electrochemical performance observed during cycling.

2. Results and Discussion

In this study, NMC 811 coin cells were electrochemically cycled against lithium metal within a potential range of 2.7 to 4.5 V. To investigate the impact of upper potential limits on the formation and efficiency of the CEI layer, two distinct cutoff voltages were employed: a standard limit at 4.3 V (referred to as the 4.3 V-limit) and an extended limit at 4.5 V (referred to as the 4.5 V-limit). Over the course of 90 cycles, the development of the CEI layer was meticulously analyzed. This involved assessing its efficiency relative to the number of cycles and conducting a comparative analysis with the efficiencies of CEI layers documented in existing literature.

In Figure 1a,g, we delineate the electrochemical characteristics of the sample. The cycling of the sample was conducted at a C-rate of C/20, interspersed with a 15 min rest period between each charging and discharging cycle. Figure 1a depicts the potential versus capacity curve across various cycles. Up to the 20th cycle, minimal polarization was observed, indicative of the sample's stability and negligible degradation. However, at the 50th cycle, a marked increase in polarization was evident relative to previous cycles. The initial cycle demonstrated a capacity of approximately 200 mAh/g, which diminished to around 143 mAh/g by the end of the battery's life cycle, culminating in a capacity retention of 71.5%. Figure 1g illustrates the progression of capacity loss and Coulombic efficiency (CE). A consistent pattern of degradation was observed. The absence of abrupt capacity loss at the graph's conclusion can be attributed to the fact that the span from 1 to 50 cycles represents merely a segment of the battery's potential state of life (SoL). This is corroborated by the CE data, which showed a modest decline of approximately 2% between the 1st and 50th cycles.

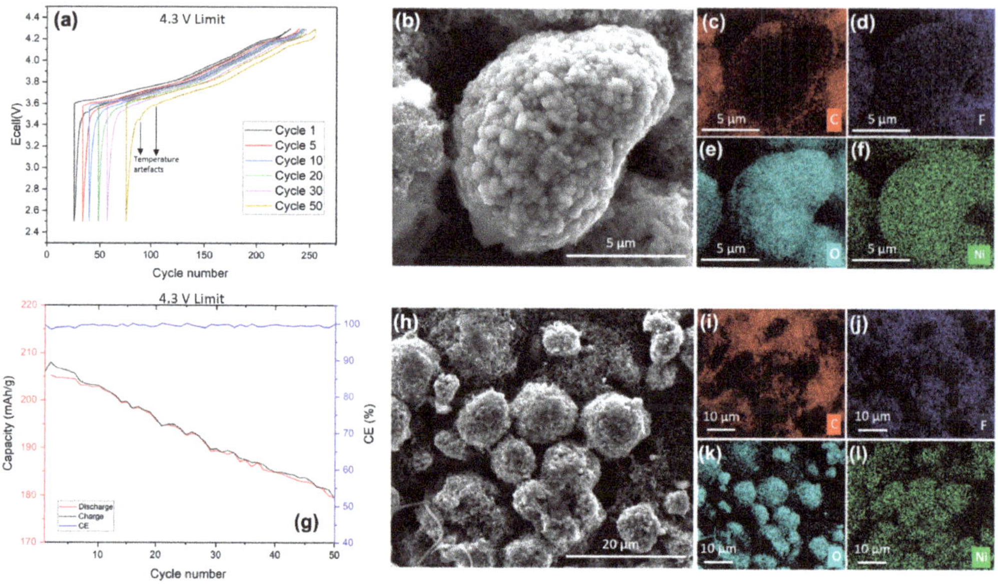

Figure 1. (**a**) Potential vs. capacity electrochemical curve of cycled NMC 811 at an upper high voltage of 4.3 V. (**b**) Secondary electron (SE) image of NMC 811 single secondary particles in the cathode sheet after cleaning (4.3 V-limit cell). EDX analysis for elemental identification of (**c**) carbon, (**d**) fluorine, (**e**) oxygen, and (**f**) nickel. (**g**) Charge and discharge lost and CE in the function of the number of cycles on the NMC 811 coin cell. (**h**) Secondary electron (SE) image of NMC 811 secondary particles in the cathode sheet after cleaning (4.3 V-limit cell). EDX analysis for elemental identification of (**i**) carbon, (**j**) fluorine, (**k**) oxygen, and (**l**) nickel.

The polarization and the variations in charge/discharge efficiency (Figure S1), along with the CE, can be primarily attributed to two distinct mechanisms leading to electrolyte degradation. These mechanisms include the formation of the CEI layer and the direct degradation of the electrolyte due to its interaction with lithium (Li) metal at the negative electrode. Li metal, especially, poses a significant risk due to its high reactivity when in direct contact with the LP30 electrolyte. This interaction results in the formation of various degradation products, such as 2,5-dioxahexanedioic acid dimethyl, CO_2, CO, and phosphates, as referenced in [14,35,36].

The cathode material was examined using SEM, and the findings are exhibited in Figure 1c–f. Images of individual secondary particles (agglomerate of primary particles) were acquired to obtain a first overview of primary particles' behavior. Analysis of both the aggregated particles (Figure 1b–f,h–l) and individual particles (Figure 1d,j) revealed the presence of fluorine on the particle surfaces. Phosphorus was absent in the spectra, indicating that the detected fluorine originates from an electrochemical process rather than being a residual salt. This observation of fluorine suggests the formation of a CEI layer, which, according to the existing literature, is likely to be a LiF layer. In Figure 1c,d,i,j, where both carbon and fluorine were observed, carbon distribution was non-uniform across the surface of the secondary particle. Conversely, fluorine exhibited a homogeneous distribution, implying the absence of carbonate compounds in the CEI layer and suggesting the presence of LiOH. However, the exact composition of the CEI layer remains undetermined due to analytical limitations. Additionally, in the Supplementary Materials (Figure S2b), an elemental analysis including manganese is provided for comparison with fluorine, considering their similar edge energies. The distinct presence of fluorine and manganese confirms that they are separate elements.

To gain deeper insight into the cell structure, analysis at the primary particle scale was conducted using STEM-EDX, with results presented in Figure 2. The high-angle annular dark-field (HAADF) imaging in Figure 2a reveals an agglomeration of primary particles, ranging in size from 0.5 to 1.5 μm. Figure 2b displays the line profile across various particle edges (indicated by a red line in Figure 2a), highlighting the presence of fluorine, particularly pronounced in the third particle examined. Figure 2c compares the Mn (Mn-Kα 5.895 keV) and F (F-K 0.676 keV) signals, showing a predominant Mn signal, with fluorine primarily detected along the edges of some particles, aligning with previous SEM-EDX observations. The C-O mapping in Figure 2d was conducted to assess the potential presence of carbonate compounds, but unlike the fluorine layer, no distinct carbon layer was observed. Finally, Figure 2e illustrates the transition metals, providing insights into the high homogeneity in composition of NMC 811 particles with no segregation.

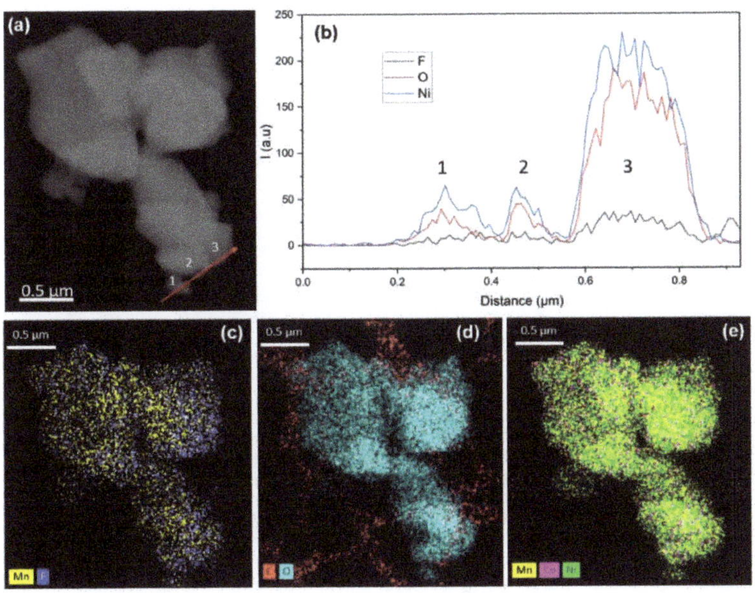

Figure 2. The 4.3 V sample: STEM-EDX analysis for primary agglomerate particles. (**a**) HAADF image. (**b**) Line profile of the selection area (red line) featuring the nickel, oxygen, and fluorine presence over it. (**c**) Color overlay featuring fluorine and manganese over agglomerate particles for LiF observation. (**d**) Color overlay featuring carbon and oxygen for carbonates' observation. (**e**) Color overlay featuring nickel, manganese, and cobalt (NMC 811).

To enhance the comprehension of the results from the STEM-EDX analysis, a comprehensive 4D-STEM investigation was conducted on two selected areas within the particle agglomerate. Crystallographic orientation and phase mapping were derived through the application of ACOM data-processing methods [37,38], coupled with the use of the ePattern suite for data noise reduction [34]. Consistent with observations from the high-angle annular dark-field (HAADF) imaging, the agglomerate was identified as a conglomerate of primary particles. ACOM analysis was specifically applied to two zones, designated as zone 1 (illustrated in Figure 3a) and zone 2 (shown in Figure 3g), where STEM-EDX data indicated a high potential for CEI layer formation.

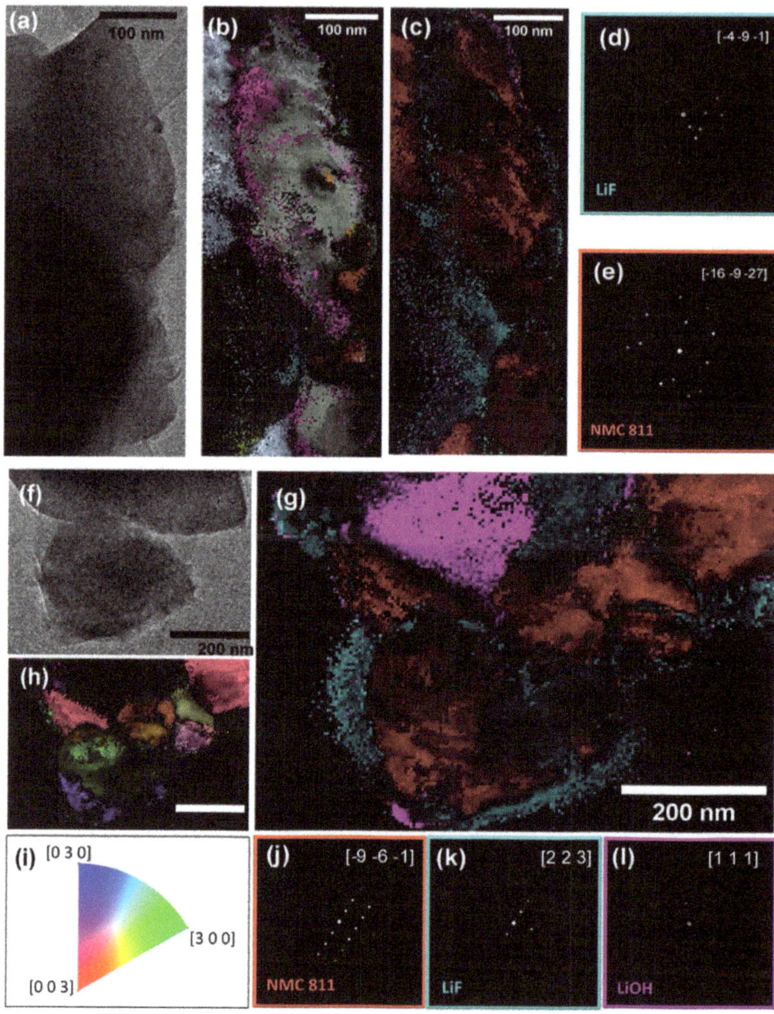

Figure 3. The 4.3 V sample: (**a**) TEM image of the 4.3 V−limit edge primary particle (zone 1). Particle edge 4D-STEM analysis: (**b**) reliability map and orientation map superposition, (**c**) phase map and phase reliability map superposition, (**d**) LiF diffraction pattern, (**e**) NMC 811 diffraction pattern, and (**f**) TEM image of the 4.3 V−limit primary particle (zone 2). 4D-STEM primary particle analysis: (**g**) reliability map and orientation map superposition, (**h**) phase map and phase reliability map superposition, (**i**) orientation color map, (**j**) NMC 811 diffraction pattern, (**k**) LiF diffraction pattern, and (**l**) LiOH diffraction pattern.

In zone 1, the crystal orientation and phase maps at a particle's edge are displayed in Figure 3b and 3c, respectively. Two primary components, NMC 811 and LiF, were identified, aligning with the STEM-EDX findings, as shown in Figure 3d,e. The orientation fidelity was notably high for the NMC 811 phase, but substantially lower for LiF. Despite this, certain areas exhibited sufficient phase and orientation reliability, affirming the presence of

LiF at the particle's edge. The diminished orientation reliability in some regions could be attributed to particle thickness or overlapping particle layers.

Regarding zone 2, the orientation mapping revealed a polycrystalline structure (Figure 3h). This finding was corroborated by TEM (Figure 3g) and STEM-HAADF (Figure 2a) imaging, which did not indicate particle superposition, thereby confirming the polycrystalline nature of the primary particles. This supports the theory of LiF layer instability due to mechanical disruption, as previously discussed. Figure 3k illustrates the uneven spatial distribution of LiF around the particle edge, mirroring observations in zone 1. Additionally, LiOH was detected in this area with high phase and orientation reliability (Figure 3l), yet it was notably absent from the particle edges.

An alternative hypothesis for the origin of the LiF component layer and LiOH involves the potential presence of water traces within the cathode material, which could facilitate the production of HF and LiF [12,39]. Despite the samples being synthesized under meticulously controlled conditions within a dry room, the likelihood of moisture contamination during the battery assembly process cannot be discounted as a contributing factor to this observed phenomenon. Furthermore, it is plausible that residual water traces are inherently present in the electrolyte's solvents [40], which may predispose the formation of a LiF layer within the CEI.

In the presented study, electrochemical characteristics of a coin cell with a 4.5 V voltage limit were elucidated, as depicted in Figure 4. This cell underwent cycling at a C/20 rate, incorporating a 15 min interlude between successive charging and discharging phases. The polarization observed in the 4.5 V-limit cell (Figure 4a) exhibited an enhanced magnitude compared to its 4.3 V-limit counterpart, yet it maintained stability throughout the initial 50 cycles.

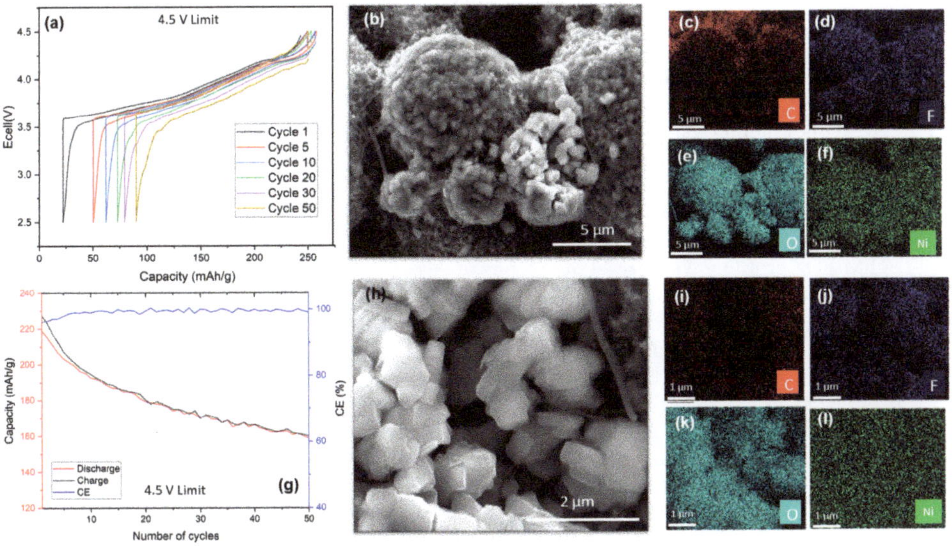

Figure 4. (**a**) Potential vs. capacity electrochemical curve of cycled NMC 811 at the voltage limit of 4.5 V. (**b**) Secondary electron (SE) image of NMC 811 secondary particles in the cathode sheet after cleaning (4.5 V-limit cell). EDX analysis for elemental identification of (**c**) carbon, (**d**) fluorine, (**e**) oxygen, and (**f**) nickel. (**g**) Charge and discharge lost and CE in the function of the number of cycles on the NMC 811 coin cell. (**h**) Secondary electron (SE) image of NMC 811 single secondary particles in the cathode sheet after cleaning (4.5 V-limit cell). EDX analysis for elemental identification of (**i**) carbon, (**j**) fluorine, (**k**) oxygen, and (**l**) nickel.

A closer examination of Figure 4b reveals a pronounced decline in capacity during the early stages of the cell's operational lifespan, stabilizing after approximately 20 cycles, similar to the pattern observed in the 4.3 V-limit sample. This initial rapid degradation can likely be ascribed to the electrolyte's accelerated deterioration under high-voltage conditions and the formation of a less stable solid electrolyte interphase (SEI) layer, which is susceptible to crystallographic and potential morphological alterations, in contrast to the 4.3 V-limit scenario [41].

Furthermore, the CE of the cell is illustrated in the same figure. While it remained relatively stable, there was an approximate 4% reduction, which is more significant than that of the 4.3 V-limit sample. It is important to note that the thermal behavior of both cells was consistent, as they were subjected to identical cycling conditions in the same environment. For the 4.5 V-limit cell, the initial capacity was approximately 230 mAh/g, diminishing to around 130 mAh/g at the end of its life cycle. This translates to a capacity retention rate of 56.5%, indicating a 21% decrease in efficiency compared to the cell cycled at a 4.3 V cutoff voltage.

The primary aim of this study was to investigate the potential presence of CEI layer constituents, particularly carbonates, through SEM-EDX analysis. Multiscale images were acquired for the same reason presented in the case of the 4.3 V-limit sample. Figure 4b illustrates a cluster of secondary particles, with at least one exhibiting fracturing. As previously noted, mechanical strain during cycling may induce deformation, potentially leading to the fracturing of secondary particles. The detection of fluorine within the interior of the particle, as shown in Figure 4f, implies pre-cycling damage. This is consistent with observations from the 4.3 V-limit sample, where fluorine was present on the surface of all analyzed particles (Figure 4e,f), indicating the formation of a reactive and potentially unstable CEI layer during cycling. In contrast, the absence of carbon on the surface of particles in the 4.5 V-limit sample suggests a lack of carbonate components. Figure S3c,d present the manganese color overlay of the 4.5 V-limit sample, demonstrating a distribution pattern distinct from that of fluorine, thus confirming the presence of the latter element and eliminating any potential misinterpretation of spectral energies.

In the scientific analysis of the 4.5 V-limit sample, a thorough STEM-EDX examination of primary particles was conducted. The findings are presented in Figure 5. This analysis entailed a meticulous study of two primary particles, each approximately 500 nm in size. Figure 5c displays a detailed mapping of manganese and fluorine elements. The presence of fluorine is evident along the edge of these particles. Consequently, a more targeted STEM-EDX profiling was carried out on one of these particles (as depicted in Figure 5a,b), revealing an inhomogeneous fluorine distribution along the edges (F-K 0.676 keV). This inhomogeneity is attributed to the varying thickness of the particles.

Parallel to the procedure executed for the 4.3 V-limit sample, a carbon-oxygen (C-O) mapping was performed. This step aimed to detect any potential carbonate layers. However, the carbon (C-K 0.278 keV) detected around the particle edges was insufficient to conclusively identify a CEI layer composed of carbonate.

Furthermore, the comparative analysis of Figure 5e, showing NMC 811 mapping, and Figure 5a, depicting STEM-HAADF imaging, provided a crucial insight. It confirmed the absence of overlap between the analyzed particles. This observation is significant, as it implies that 4D-STEM analysis could be effectively employed to ascertain the polycrystalline nature of these primary particles without any complications arising from particle superposition.

The 4D-STEM ACOM analysis was performed over the zone of interest, as shown in Figure 6. First, the orientation map (Figure 6b) showed a polycrystalline composition for the particle on the left of the scan. For the particle on the right, the thickness did not allow to have a high enough orientation/phase reliability to conclude on a polycrystalline or monocrystalline composition. This information allowed us to reaffirm the hypothesis of lower performance of the 4.5 V-limit sample in comparison to the 4.3 V-limit due to layer rupture contributing to the deformation at a higher voltage. In Figure 6d, the phase map shows the presence of both NMC 811 (Figure 6f) and LiF (Figure 6e), and the orientation and phase maps (together with STEM-EDX analysis) confirm the presence of LiF on the particle edge, with higher confidence than the 4.3 V-limit sample. The higher confidence could be because at a higher voltage, the more aggressive degradation of the electrolyte leads to more LiF formation. However, this does not mean better protection from capacity loss, as we could expect from the reviewed literature. On the contrary, it rather represents a higher loss due to the non-stability of the CEI.

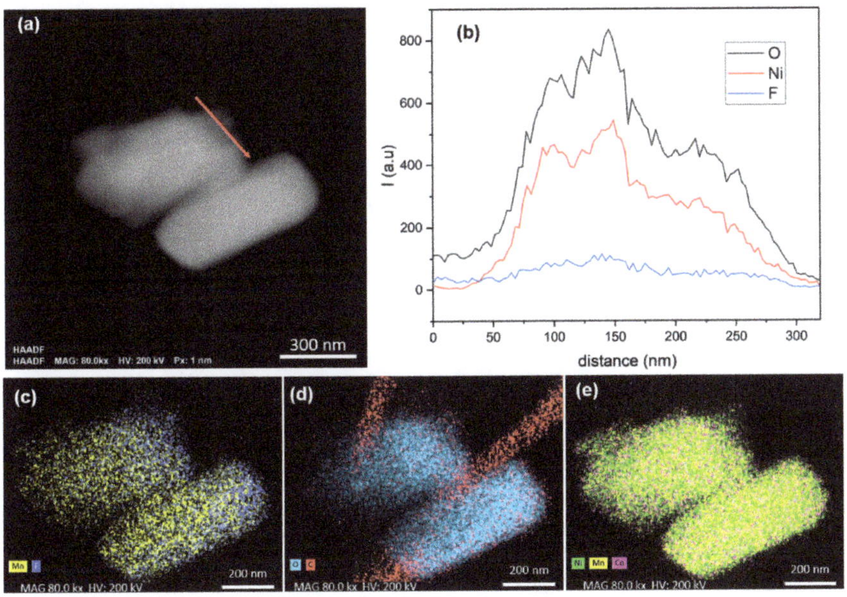

Figure 5. The 4.5 V sample: STEM-EDX analysis for primary particles. (**a**) HAADF image. (**b**) Line profile of the selection area (red arrow (**a**)), featuring the nickel, oxygen, and fluorine presence over it. (**c**) Color overlay featuring fluorine and manganese over particles for LiF observation. (**d**) Color overlay featuring carbon and oxygen for carbonates' observation. (**e**) Color overlay featuring nickel, manganese, and cobalt (NMC 811).

Considering the analysis of both samples, the LiF layer was not formed homogeneously over the entire particle edge. A possible explanation for this is the fact that the CEI formation occurs at the surface of secondary particles, meaning that during grinding, core particles do not present any CEI formation. Moreover, the CEI layer can be affected due to the mechanical energy induced by the grinding of the particles.

Li inorganic compounds were expected to be present in the CEI layer, as has already been reported in the literature. Carbonates and oxides were completely absent, LiOH was found in the sample cycled at 4.3 V in a small quantity, and LiF was present in both samples (4.3 V and 4.5 V) as the predominant component. Even though LiF has been reported to be used as a CEI protective layer against degradation in NMC 811, as we presented in the introduction of this paper, in our case, the samples still presented a retention capacity between 70% and 50%, and these values are close to those found in other works for NMC 811 samples with any treatment or additive against degradation.

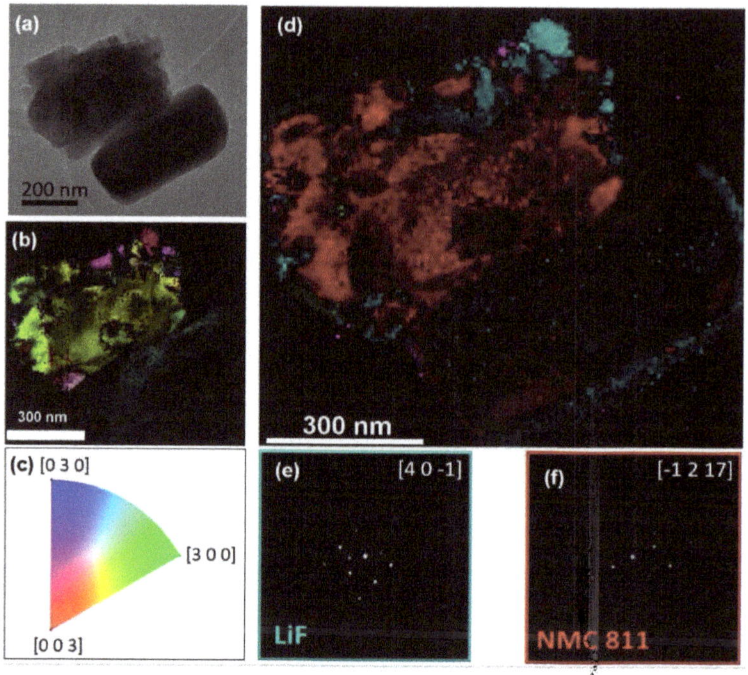

Figure 6. The 4.5 V sample. (**a**) TEM image of 4.5 V-limit primary particles, (**b**) 4D-STEM analysis of primary particles: reliability map and orientation map superposition, (**c**) orientation color map, (**d**) 4D-STEM analysis of primary particles: phase map and phase reliability map superposition, (**e**) diffraction pattern (DF) for LIF (blue) component, and (**f**) DF for NMC 811 component (red).

3. Conclusions

Ni-rich cathode materials are known for the several degradation mechanisms present due to their high potential working conditions. Chemical and mechanical degradation have been reported in numerous works. In this research, we explored the complex degradation mechanisms in Ni-rich cathode materials, known for their susceptibility to chemical and mechanical degradation under high-potential conditions. Our primary focus was to investigate strategies for mitigating these degradation processes without compromising the capacity of Ni-rich materials. A pivotal aspect of this endeavor involved the formation of a stable CEI layer.

Utilizing advanced characterization techniques, such as 4D-STEM with automated crystal orientation mapping, along with electrochemical analysis, SEM-EDX, and STEM-EDX, we conducted an in-depth study of CEI layer formation in conventional NMC 811 coin cell configurations. These configurations included a lithium metal anode and an LP30 electrolyte, examined at both cutoff voltage limits and beyond.

Our findings revealed the formation of an LiF-based CEI layer in both samples, potentially resulting from water traces in the cathode during preparation. While LiF is frequently lauded for its potential in mitigating degradation phenomena in battery systems, our observations indicated that this layer does not form prior to cycling, and its protective efficacy is not consistent with results reported by other research groups. Our electrochemical data further highlighted a more pronounced degradation at higher cutoff voltages (4.5 V-limit), which could be attributed to the continuous formation and dissolution of the CEI layer. This process, coupled with particle deformation, exacerbates the degradation of the electrolyte at elevated voltages.

Our methodological approach and experimental setup proved effective in studying and understanding CEI formation. However, it is critical to note that CEI formation is predominantly a surface phenomenon at the secondary particle scale. This implies a reduced likelihood of observing primary particles with a CEI layer in TEM analysis, especially considering that sample preparation involves powder grinding, which may induce postmortem degradation of the existing CEI layer.

In perspective, for a better comprehensive analysis of CEI formation, liquid electrochemical in situ TEM cycling analysis techniques, including STEM-EDX and 4D-STEM, are indispensable. Our study also highlighted the inevitability of parasitic reactions, even in controlled environments, across both homemade and commercial batteries. These insights suggest that future advancements in battery performance may hinge on integrating electrolyte additives or adopting cathode material coatings, as explored in previous studies. This direction holds promise for enhancing the durability and efficiency of Ni-rich cathode materials in high-potential battery applications. Finally, it is imperative to conduct a more advanced examination of phase transformations occurring at the surface of Ni-rich NMC in relation to the applied electrochemical potential and the composition of the CEI layer. This investigation is crucial for advancing our understanding of the interplay between CEI layer formation and the alteration in crystallinity of NMC materials.

4. Materials and Methods

4.1. TEM/STEM Analysis

In the context of TEM analysis, the powder from the cathode sheet was detached from the current collector, manually pulverized, and subsequently stored in an aerated glass flask. HAADF-STEM imaging was conducted using a Tecnai G2 F20 S-Twin (Thermo Fisher Scientific, Waltham, MA, USA) system, operated at an accelerating voltage of 200 kV and equipped with a C2 aperture of 70 μm. Additionally, for the acquisition of elemental maps, an energy-dispersive X-ray spectroscope (EDX, Xflash, Bruker, Berlin, Germany) was employed in STEM mode. EDX-STEM analysis was performed in drifted corrected mode, with an EDS map interval of 10 s, dexell time of 8 μs, and a stop mode relative value of 50%. For Q-mapping, the TEM Cliff Lorimer B method was used, where the pixel size was 2 nm, time per pixel was 64 μs, and total scanning time was 532 s.

4.2. 4D-STEM ACOM Analysis

The TEM investigations were conducted using an accelerating voltage set to 200 kV. During the diffraction experiments, the camera length was meticulously maintained at 300 mm. A precesion angle of 0.7° was employed, aimed at minimizing dynamical scattering effects. The condenser aperture was precisely configured to 10 μm, resulting in a convergence semi-angle of 0.4 mrad. For electron beam control, Gun lens 3 was utilized, with the spot size adjusted to 5. The electron dosage for 4D-STEM analysis was established at 150 $e/\text{Å}^2/s$. Data processing of the 4D-STEM dataset was executed utilizing the ePattern suite software, which facilitated denoising operations with a prominence value set at 5. Subsequently, the ASTAR software package (developed by Nanomegas, Brussels, Belgium) was applied for the reconstruction of phase and orientation maps. This was achieved through the Automated Crystal Orientation Mapping (ACOM) technique, which relies on a pattern-matching algorithm. The resolution of each diffraction pattern was configured to 512 × 512 pixels. Acquisition of these diffraction patterns was carried out using a OneView CMOS camera, manufactured by Gatan, Pleasanton, CA, USA.

4.3. SEM Analysis

An environmental SEM (ESEM), specifically the FEI Quanta 200 Field Emission Gun (FEG) model, augmented with an advanced energy-dispersive X-ray (EDX) microanalyzer (X-Max 80, Oxford Instruments Co., Abingdon, UK), was employed for detailed microstructural analysis. Imaging modalities, including secondary electron (SE) and backscattered electron (BSE) techniques, were utilized, conducted under a controlled high-vacuum environment at an electron acceleration voltage range of 10 to 15 keV. EDX spectroscopy was consistently executed at an acceleration voltage of 15 keV to ensure optimal elemental characterization. For SEM imaging, the cathode sheet was subjected to examination in its post-recovery state, following a meticulous cleaning process within a controlled atmosphere glovebox. The cathode sheet was carefully placed over the SEM sample holder using carbon tape, with no further manipulation.

4.4. Electrochemical Analysis

The positive electrode material, namely NMC 811, was procured from NEI Corporation (Piscataway, NJ, USA) in the form of cathode sheets, each measuring 127 mm by 254 mm, and adhered to an aluminum current collector with a thickness of 16 µm. The composition of the cathode involved a blend of NMC 811 (constituting 90% of the active material), polyvinylidene fluoride (PVDF) as a binder (5%), and Super P conductive carbon (5%), achieving a uniform thickness of approximately 60 µm ± 5%. For the electrolyte, 100 µL of LP30, a commonly utilized formulation comprising lithium hexafluorophosphate ($LiPF_6$) in a binary solvent system of ethylene carbonate (EC) and dimethyl carbonate (DMC) in a 1:1 volumetric ratio, was employed. A 1 mm-thick fiberglass separator was integrated into the design. Lithium metal, serving as the negative electrode, was sourced from Sigma Aldrich (Saint Louis, MO, USA), preserved in a controlled dry environment, and prepared as a thin foil for subsequent processing.

The coin cell assembly was conducted within a dry room environment, utilizing a 13 mm punched cathode, a 15 mm punched separator, and an 8 mm punched lithium metal. Post-cycling, the coin cell was disassembled within a nitrogen-filled glovebox, and the cathode sheet was subjected to a thorough cleaning process using DMC. This cleaning involved a three-cycle spinning protocol, each at 3600 rpm for 3 min, followed by a drying phase at 80 °C for one hour in an air atmosphere. Electrochemical characterization was performed by cycling the coin cell using a BSC-COM Biologic Potentiostat, with the operational parameters and data acquisition managed by EC-Lab software, version 11.34.

Supplementary Materials: The following supporting information can be downloaded at: https://www.mdpi.com/article/10.3390/sym16030301/s1, Figure S1: Potential vs. capacity electrochemical curve of cycled NMC 811 at an upper high voltage of 4.3 V; Figure S2: Secondary electron (SE) image of NMC secondary particles in cathode sheet after cleaning from 4.3V-limit cell; Figure S3: Secondary electron (SE) image of NMC secondary particles in cathode sheet after cleaning from 4.5V-limit cell. Supporting information shows SEM-EDX images of secondary particles featuring Mn and F. The codes of cif files used in 4D-STEM and more details of the acquisition conditions are presented as well.

Author Contributions: Conceptualization, A.D.; methodology, A.D. and K.G.-M.; software, N.F. and J.J.; validation, A.D. and A.J.; formal analysis, K.G.-M. and J.J.; investigation, K.G.-M. and J.J.; writing—original draft preparation, K.G.-M. and A.D.; writing—review and editing, A.D. and A.J.; supervision, A.D.; project administration, A.D.; funding acquisition, A.D. All authors have read and agreed to the published version of the manuscript.

Funding: As a part of the DESTINY PhD program, this publication is acknowledged by funding from the European Union's Horizon 2020 research and innovation program under the Marie Skłodowska-Curie Actions COFUND (Grant Agreement #945357). A part of the funding has been provided by the French Research Agency (ANR) as part of the DestiNa-ion Operando project (ANR-19-CE42-0014).

Data Availability Statement: For any data requirement, please contact the corresponding author.

Acknowledgments: The UPJV and RS2E electron microscopy platforms were utilized for this research.

Conflicts of Interest: The authors declare no conflicts of interest.

References

1. Zhao, Y.; Pohl, O.; Bhatt, A.I.; Collis, G.E.; Mahon, P.J.; Rüther, T.; Hollenkamp, A.F. A Review on Battery Market Trends, Second-Life Reuse, and Recycling. *Sustain. Chem.* **2021**, *2*, 167–205. [CrossRef]
2. A Vision for a Sustainable Battery Value Chain in 2030 Report: Unlocking the Full Potential to Power Sustainable Development and Climate Change Mitigation. Available online: https://www3.weforum.org/docs/WEF_A_Vision_for_a_Sustainable_Battery_Value_Chain_in_2030_Report.pdf (accessed on 26 September 2023).
3. Sun, H.; Zhao, K. Electronic Structure and Comparative Properties of LiNixMnyCozO2 Cathode Materials. *J. Phys. Chem. C* **2017**, *121*, 6002–6010. [CrossRef]
4. Li, J.; Downie, L.E.; Ma, L.; Qiu, W.; Dahn, J.R. Study of the Failure Mechanisms of $LiNi_{0.8}Mn_{0.1}Co_{0.1}O_2$ Cathode Material for Lithium Ion Batteries. *J. Electrochem. Soc.* **2015**, *162*, A1401–A1408. [CrossRef]
5. Takacova, Z.; Havlik, T.; Kukurugya, F.; Orac, D. Cobalt and Lithium Recovery from Active Mass of Spent Li-Ion Batteries: Theoretical and Experimental Approach. *Hydrometallurgy* **2016**, *163*, 9–17. [CrossRef]
6. Jung, R.; Metzger, M.; Maglia, F.; Stinner, C.; Gasteiger, H.A. Chemical versus Electrochemical Electrolyte Oxidation on NMC111, NMC622, NMC811, LNMO, and Conductive Carbon. *J. Phys. Chem. Lett.* **2017**, *8*, 4820–4825. [CrossRef]
7. Jung, R.; Metzger, M.; Maglia, F.; Stinner, C.; Gasteiger, H. Oxygen Release and Its Effect on the Cycling Stability of $LiNi_xMn_yCo_zO_2$ (NMC) Cathode Materials for Li-Ion Batteries. *J. Electrochem. Soc.* **2017**, *164*, A1361–A1377. [CrossRef]
8. Noh, H.-J.; Youn, S.; Yoon, C.; Sun, Y.-K. Comparison of the Structural and Electrochemical. Properties of Layered $Li[Ni_xCo_yMn_z]O_2$ (x = 1/3, 0.5, 0.6, 0.7, 0.8 and 0.85) Cathode Material for Lithium-Ion Batteries. *J. Power Sources* **2013**, *233*, 121–130. [CrossRef]
9. Cherkashinin, G.; Motzko, M.; Schulz, N.; Späth, T.; Jaegermann, W. Electron Spectroscopy Study of $Li[Ni,Co,Mn]O_2$/Electrolyte Interface: Electronic Structure, Interface Composition, and Device Implications. *Chem. Mater.* **2015**, *27*, 2875–2887. [CrossRef]
10. Takahashi, I.; Kiuchi, H.; Ohma, A.; Fukunaga, T.; Matsubara, E. Cathode Electrolyte Interphase Formation and Electrolyte Oxidation Mechanism for Ni-Rich Cathode Materials. *J. Phys. Chem. C* **2020**, *124*, 9243–9248. [CrossRef]
11. Gauthier, M.; Carney, T.J.; Grimaud, A.; Giordano, L.; Pour, N.; Chang, H.-H.; Fenning, D.P.; Lux, S.F.; Paschos, O.; Bauer, C.; et al. Electrode–Electrolyte Interface in Li-Ion Batteries: Current Understanding and New Insights. *J. Phys. Chem. Lett.* **2015**, *6*, 4653–4672. [CrossRef]
12. Thapaliya, B.P.; Misra, S.; Yang, S.; Jafta, C.J.; Meyer III, H.M.; Bagri, P.; Unocic, R.R.; Bridges, C.A.; Dai, S. Enhancing Cycling Stability and Capacity Retention of NMC811 Cathodes by Reengineering Interfaces via Electrochemical Fluorination. *Adv. Mater. Interfaces* **2022**, *9*, 2200035. [CrossRef]
13. Azcarate, I.; Yin, W.; Méthivier, C.; Ribot, F.; Laberty-Robert, C.; Grimaud, A. Assessing the Oxidation Behavior of EC:DMC Based Electrolyte on Non-Catalytically Active Surface. *J. Electrochem. Soc.* **2020**, *167*, 080530. [CrossRef]
14. Gireaud, L.; Grugeon, S.; Pilard, S.; Guenot, P.; Tarascon, J.-M.; Laruelle, S. Mass Spectrometry. Investigations on Electrolyte Degradation Products for the Development of Nanocomposite Electrodes in Lithium Ion Batteries. *Anal. Chem.* **2006**, *78*, 3688–3698. [CrossRef]
15. Wu, F.; Fang, S.; Kuenzel, M.; Mullaliu, A.; Kim, J.-K.; Gao, X.; Diemant, T.; Kim, G.-T.; Passerini, S. Dual-Anion Ionic Liquid Electrolyte Enables Stable Ni-Rich Cathodes in Lithium-Metal Batteries. *Joule* **2021**, *5*, 2177–2194. [CrossRef]
16. Heist, A.; Hafner, S.; Lee, S.-H. High-Energy Nickel-Rich Layered Cathode Stabilized by Ionic Liquid Electrolyte. *J. Electrochem. Soc.* **2019**, *166*, A873. [CrossRef]
17. Xu, R.; Sun, H.; de Vasconcelos, L.S.; Zhao, K. Mechanical and Structural Degradation of $LiNi_xMn_yCo_zO_2$ Cathode in Li-Ion Batteries: An Experimental Study. *J. Electrochem. Soc.* **2017**, *164*, A3333–A3341. [CrossRef]
18. Sungjemmenla, S.K.V.; Soni, C.B.; Kumar, V.; Seh, Z.W. Understanding the Cathode–Electrolyte Interphase in Lithium-Ion Batteries. *Energy Technol.* **2022**, *10*, 2200421. [CrossRef]
19. Maleki Kheimeh Sari, H.; Li, X. Controllable Cathode–Electrolyte Interface of $Li[Ni_{0.8}Co_{0.1}Mn_{0.1}]O_2$ for Lithium Ion Batteries: A Review. *Adv. Energy Mater.* **2019**, *9*, 1901597. [CrossRef]
20. Jiang, S.; Xu, X.; Yin, J.; Lei, Y.; Guan, H.; Gao, Y. High-Performance $Li/LiNi_{0.8}Co_{0.1}Mn_{0.1}O_2$ Batteries Enabled by Optimizing Carbonate-Based Electrolyte and Electrode Interphases via Triallylamine Additive. *J. Colloid Interface Sci.* **2023**, *644*, 415–425. [CrossRef] [PubMed]
21. Chen, Z.; Zhang, H.; Xu, H.; Dong, S.; Jiang, M.; Li, Z.; Cui, G. In Situ Generated Polymer Electrolyte Coating-Based Janus Interfaces for Long-Life LAGP-Based NMC811/Li Metal Batteries. *Chem. Eng. J.* **2022**, *433*, 133589. [CrossRef]
22. Beltrop, K.; Klein, S.; Nölle, R.; Wilken, A.; Lee, J.J.; Köster, T.K.J.; Reiter, J.; Tao, L.; Liang, C.; Winter, M.; et al. Triphenylphosphine Oxide as Highly Effective Electrolyte Additive for Graphite/NMC811 Lithium Ion Cells. *Chem. Mater.* **2018**, *30*, 2726–2741. [CrossRef]
23. Jing, Y.; Shao, Y.; Xin, H.L. On The Efficacy of Cobalt Boride Coating on NMC-811 Cathode Under Vinylene Carbonate Additive, High Temperature and Air Shelving Conditions. *J. Electrochem. Soc.* **2023**, *170*, 010519. [CrossRef]
24. Ji, H.; Urban, A.; Kitchaev, D.A.; Kwon, D.-H.; Artrith, N.; Ophus, C.; Huang, W.; Cai, Z.; Shi, T.; Kim, J.C.; et al. Hidden Structural and Chemical Order Controls Lithium Transport in Cation-Disordered Oxides for Rechargeable Batteries. *Nat. Commun.* **2019**, *10*, 592. [CrossRef] [PubMed]

25. Folastre, N.; Cherednichenko, K.; Cadiou, F.; Bugnet, M.; Rauch, E.; Jacob, O.; Croguennec, L.; Masquelier, C.; Demortiere, A. Multimodal Study of Dis-Sodiation Mechanisms within Individual $Na_3V_2(PO_4)_2F_3$ Cathode Crystals Using 4D-STEM-ASTAR and STXM-XANES. *Microsc. Microanal.* **2021**, *27*, 3446–3447. [CrossRef]
26. Su, Z.; De Andrade, V.; Cretu, S.; Yin, Y.; Wojcik, M.J.; Franco, A.A.; Demortière, A. X-Ray Nanocomputed Tomography in Zernike Phase Contrast for Studying 3D Morphology of Li–O_2 Battery Electrode. *ACS Appl. Energy Mater.* **2020**, *3*, 4093–4102. [CrossRef]
27. Cherednichenko, K.; Antitomaso, P.; Rabuel, F.; Folastre, N.; Masquelier, C.; Croguennec, L.; Veron, M.; Rauch, E.; Demortière, A. Study of (Dis)-Sodiation Mechanisms within Individual $Na_3V_2(PO_4)_2F_3$ Cathode Crystals By 4D-STEM Phase and Orientation Mapping Using ASTAR System. *Meet. Abstr.* **2020**, *MA2020-01*, 88. [CrossRef]
28. Ophus, C. Four-Dimensional Scanning Transmission Electron Microscopy (4D-STEM): From Scanning Nanodiffraction to Ptychography and Beyond. *Microsc. Microanal.* **2019**, *25*, 563–582. [CrossRef] [PubMed]
29. Thomas, J.M.; Leary, R.K.; Eggeman, A.S.; Midgley, P.A. The Rapidly Changing Face of Electron Microscopy. *Chem. Phys. Lett.* **2015**, *631–632*, 103–113. [CrossRef]
30. Da Silva, B.C.; Sadre Momtaz, Z.; Monroy, E.; Okuno, H.; Rouviere, J.-L.; Cooper, D.; Den Hertog, M.I. Assessment of Active Dopants and p–n Junction Abruptness Using In Situ Biased 4D-STEM. *Nano Lett.* **2022**, *22*, 9544–9550. [CrossRef]
31. Xu, J. Critical Review on Cathode–Electrolyte Interphase Toward High-Voltage Cathodes for Li-Ion Batteries. *Nano-Micro Lett.* **2022**, *14*, 166. [CrossRef]
32. Dhiman, R. Investigation of Cathode Electrolyte Interphase Layer in V2O5 Li-Ion Battery Cathodes: Time and Potential Effects. *J. Electrochem. Soc.* **2021**, *168*, 040512. [CrossRef]
33. Bhatia, A.; Cretu, S.; Hallot, M.; Folastre, N.; Berthe, M.; Troadec, D.; Roussel, P.; Pereira-Ramos, J.-P.; Baddour-Hadjean, R.; Lethien, C.; et al. In Situ Liquid Electrochemical TEM Investigation of $LiMn_{1.5}Ni_{0.5}O_4$ Thin Film Cathode for Micro-Battery Applications. *Small Methods* **2022**, *6*, 2100891. [CrossRef] [PubMed]
34. Folastre, N.; Cao, J.; Oney, G.; Park, S.; Jamali, A.; Masquelier, C.; Croguennec, L.; Veron, M.; Rauch, E.; Demortiere, A. Adaptive Diffraction Image Registration for 4D-STEM to Optimize ACOM Pattern Matching. 2023. Available online: https://doi.org/10.21203/rs.3.rs-2942207/v1 (accessed on 4 October 2023).
35. Yoshida, H.; Fukunaga, T.; Hazama, T.; Terasaki, M.; Mizutani, M.; Yamachi, M. Degradation Mechanism of Alkyl Carbonate Solvents Used in Lithium-Ion Cells during Initial Charging. *J. Power Sources* **1997**, *68*, 311–315. [CrossRef]
36. Brand, M.; Gläser, S.; Geder, J.; Menacher, S.; Obpacher, S.; Jossen, A.; Quinger, D. Electrical Safety of Commercial Li-Ion Cells Based on NMC and NCA Technology Compared to LFP Technology. *World Electr. Veh. J.* **2013**, *6*, 572–580. [CrossRef]
37. Rauch, E.F.; Véron, M. Automated Crystal Orientation and Phase Mapping in TEM. *Mater. Charact.* **2014**, *98*, 1–9. [CrossRef]
38. Rauch, E.; Portillo, J.; Nicolopoulos, S.; Bultreys, D.; Rouvimov, S.; Moeck, P. Automated Nanocrystal Orientation and Phase Mapping in the Transmission Electron Microscope on the Basis of Precession Electron Diffraction. *Z. Für Krist.* **2010**, *225*, 103–109. [CrossRef]
39. Zhai, P.; Liu, L.; Gu, X.; Wang, T.; Gong, Y. Interface Engineering for Lithium Metal Anodes in Liquid Electrolyte. *Adv. Energy Mater.* **2020**, *10*, 2001257. [CrossRef]
40. Strmcnik, D.; Castelli, I.E.; Connell, J.G.; Haering, D.; Zorko, M.; Martins, P.; Lopes, P.P.; Genorio, B.; Østergaard, T.; Gasteiger, H.A.; et al. Electrocatalytic Transformation of HF Impurity to H_2 and LiF in Lithium-Ion Batteries. *Nat. Catal.* **2018**, *1*, 255–262. [CrossRef]
41. Wang, L.; Su, Q.; Han, B.; Shi, W.; Du, G.; Wang, Y.; Li, H.; Gu, L.; Zhao, W.; Ding, S.; et al. Unraveling the Degradation Mechanism of $LiNi_{0.8}Co_{0.1}Mn_{0.1}O_2$ at the High Cutoff Voltage for Lithium Ion Batteries. *J. Energy Chem.* **2023**, *77*, 428–437. [CrossRef]

symmetry MDPI

Article

Determination of Na^+ Cation Locations in Nanozeolite ECR-1 Using a 3D ED Method

Taylan Örs [1,2], Irena Deroche [1,2], Corentin Chatelard [3,4], Mathias Dodin [3], Raquel Martinez-Franco [3], Alain Tuel [4] and Jean-Louis Paillaud [1,2,*]

1 Institut de Science des Matériaux de Mulhouse (IS2M), UMR CNRS 7361, Université de Haute-Alsace, 15 Rue Jean Starcky, 68100 Mulhouse, France; taylan.ors@uha.fr (T.Ö.); irena.deroche@uha.fr (I.D.)
2 Université de Strasbourg, 67000 Strasbourg, France
3 IFP Energies Nouvelles-Etablissement de Lyon, Rond-Point de l'Échangeur de Solaize, BP 3, 69360 Solaize, France; corentin.chatelard@orange.fr (C.C.); mathias.dodin@ifpen.fr (M.D.); raquel.martinez-franco@ifpen.fr (R.M.-F.)
4 Institut de Recherches sur la Catalyse et l'Environnement de Lyon (IRCELYON), UMR CNRS 5256, Université Lyon 1, 2 Avenue Einstein, 69626 CEDEX Villeurbanne, France; alain.tuel@ircelyon.univ-lyon1.fr
* Correspondence: jean-louis.paillaud@uha.fr

Abstract: Until now, the comprehensive structural analysis of single crystals of zeolite ECR-1, an aluminosilicate with the **EON** topology, has been hindered owing to the submicron dimensions of the obtained crystals. Additionally, this zeolite, which is characterized by a topology comprising alternating periodic building units of **MAZ** and **MOR** layers, exhibits stacking faults that impede accurate refinement through the Rietveld method. In this report, we present, for the first time, the structure of ECR-1 elucidated by studying a nanocrystal with a significantly reduced number of stacking faults. The sample used was synthesized hydrothermally using trioxane as the organic structure-directing agent. The structure determination was conducted using precession electron diffraction (PED) at 103 K. Partial dehydration occurred owing to the high vacuum conditions in the TEM sample chamber. From the dynamical refinement (R_{obs} = 0.097), 8.16 Na^+ compensating cations were localized on six distinct crystallographic sites, along with approximately four water molecules per unit cell. Furthermore, a canonical Monte Carlo computational study was conducted to compare the experimental cationic distribution and location of water molecules with the simulation.

Keywords: nanozeolite; ECR-1; 3D ED; precession; kinematical and dynamical refinement; Monte Carlo simulation

1. Introduction

Zeolites, which are microporous aluminosilicate materials, have extensive applications in adsorption, catalysis, and cation exchange for water treatment [1,2]. The overall negative electrical charge of zeolite frameworks allows them to accommodate charge-compensating cations in their pores and cages. The locations and natures of these cations significantly influence their physicochemical properties. For instance, faujasite-type zeolites with **FAU** topology [3] can be used in various industrial processes, depending on their Si/Al ratios and the nature of the compensating cations [4]. A notable example is lithium-exchanged LSX (Low Silica X), a faujasite with a Si/Al ratio of one, serving as the primary adsorbent for oxygen production from air [5]. Faujasites with higher Si/Al ratios, such as Y zeolites, also play crucial roles as catalysts, with key applications in cracking, isomerization, and hydrocarbon synthesis. In particular, in different enhanced forms, zeolite Y serves as the cracking component in FCC (fluid catalytic cracking) catalysts [6]. Many zeolites with distinct topologies have been synthesized to meet the ever-growing demands of industrial processes. ECR-1, discovered in 1987 [7], is one example. Zeolites with **EON** topology feature a 2-dimensional pore system defined by straight 12-membered ring (12-MR) channels interconnected through 8-membered ring (8-MR) openings. As illustrated in Figure 1, this topology can be viewed as a strict alternation of periodic building units (PerBU), namely **MAZ** (PerBU1) and **MOR** (PerBU2) layers. PerBU1 comprises columns along the *a*-axis of *gme* composite building units (CBUs) that are linked to each other via 6-MRs [8]. ECR-1 can be synthesized using various organic structure-directing agents (OSDAs), including bis-(2-hydroxyethyl)- or bis-(2-hydroxypropyl)-dimethylammonium cations [7], adamantane-containing diquaternary alkylammonium iodides [9], and tetramethylammonium [10]. It can even be produced without an organic molecule, thereby reducing production costs [11,12]. ECR-1 has been also produced efficiently via a radicalized seeds-assist route [12]. ECR-1 holds potential applications in the petroleum industry [12] and seawater desalination [13].

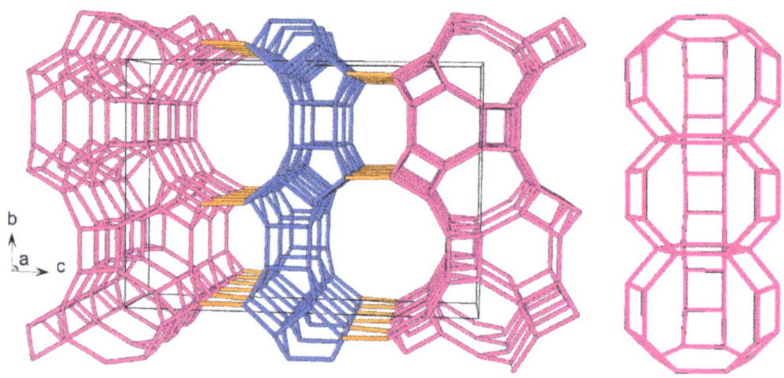

Figure 1. EON topology (**left**) along the *a*-axis, with PerBU1 and PerBU2 shown in purple and blue, respectively. Perpendicular view (**right**) illustrates a column formed by stacking three *gme* CBUs. These columns are interconnected through 8-MRs, resulting in the characteristic wavy appearance of PerBU1.

The crystal structure of ECR-1 was initially modeled based on high-resolution transmission electron microscopy (HRTEM) imaging [14]. Subsequent refinement from synchrotron powder diffraction data using the Rietveld method confirmed a structure consisting of alternating mazzite and mordenite layers with a 1:1 stacking sequence ratio [10]. However, variations in synthesis conditions and crystal growth kinetics led to different stacking fault densities. Cation locations were determined during the same Rietveld study [10] conducted on the hydrated form and under ambient conditions. Despite the use of multiple synthesis methods, no structural study from diffraction techniques on single crystal ECR-1 could be carried out owing to the submicron scale of the crystals obtained.

In a recent study, we explored the use of small, six-ring cyclic ether molecules as the organic structure-directing agent (OSDA) to synthesize omega (a synthetic analog of the zeolite mazzite) and ECR-1 zeolites [15]. Notably, 1,3,5-trioxane was employed to synthesize ECR-1. The resulting nanoscale crystals were too small for proper structural analysis from single-crystal X-ray diffraction. Moreover, in the case of ECR-1, diffraction-line broadening owing to the nanosized crystals and overlapping and stacking faults is highly problematic for proper structural analysis using the Rietveld method. For this reason, 3D electron diffraction (3D ED), which is an ensemble of methods in continuous development for the last 20 years, has been proposed for the structural analysis of nanosized single crystals [16]. Indeed, conventional electron diffraction on transmission electron microscopes (TEM) is well established for the characterization of nanoscale materials. However, dynamical scattering observed in electron diffraction does not allow conventional structure analysis to be employed routinely for X-ray diffraction data. To address this, emerging techniques like precession electron diffraction (PED) [17,18], automated diffraction tomography (ADT) [19], rotation electron diffraction (RED) [20], continuous rotation electron diffraction (cRED), and serial rotation electron diffraction (SerialRED) [21] aim to reduce the dynamic effects on observed intensities. Several reviews describe these 3D ED methods and their recent evolutions [21–27]. These methods have proved highly effective in resolving the structures of many zeolites. They have also been commonly used in combination with X-ray powder diffraction data and/or high-resolution transmission electron microscopy (HRTEM) to elucidate the structure of many zeolitic materials with varying degrees of disorder, such as IM-17, IM-18, MZS-1, zeolite beta, and RUB-5 [28–32].

The objective of the present study was to solve the structure of ECR-1 from a nanosized single crystal (with significantly reduced stacking faults) and to locate the positions of the sodium compensation cations using a 3D ED method (PED), which was chosen owing to its advantages over powder diffraction.

2. Materials and Methods

The ECR-1 of topology **EON** [3] used in this study was obtained using hydrothermal synthesis following the protocol described by Chatelard et al. [15]. In brief, a gel composed of 10 SiO_2–Al_2O_3–7.5 trioxane–1.7 Na_2O–140 H_2O was stirred at ambient temperature for 3 h before being placed in an autoclave, which was then fixed in an oven equipped with a rotating mechanism for crystallization at 115 °C for 7 days under dynamic conditions (60 rpm). Scanning electron microscopy on the as-synthesized sample showed that the powder was composed of bundles of more or less elongated needle-like crystals of a few hundred nm long and varying diameters [15].

Energy dispersive X-ray spectroscopy (EDX) analysis was performed at a voltage of 15 kV using a JSM-7900F scanning electron microscope (SEM) (JEOL, Tokyo, Japan) equipped with a Quantax analyzer system (BRUKER, Berlin, Germany) made of two XFlash 6-30 detectors. Prior to analysis, the sample was coated with a thin film of carbon using a Baltec SCD004 coater (Balzers, Bal Tec AG, Fürstentum, Lichtenstein).

The high-resolution synchrotron powder-diffraction pattern (SPDP) of the ECR-1 sample was obtained at the ID22 beamline of the ESRF in Grenoble, France [33]. The setup involved a bank of nine scintillation detectors, each preceded by nine Si 111 analyzer crystals, using vertical scanning to measure the diffracted intensity as a function of 2θ, with the channels nominally spaced 2° apart. This Debye–Scherrer capillary geometry configuration guarantees an instrumental contribution to the full width at half-maximum (FWHM) of the diffracted peaks, typically around 0.003° (2θ) in optimal conditions. ECR-1 powder was placed in a thin-walled glass capillary (1 mm in diameter) mounted on a spinner on the axis of the powder diffractometer. A Le Bail refinement [34] was conducted using the free software suite GSAS-II [35].

The PED measurements were conducted using a CM200 TEM (Philips, Eindhoven, The Netherlands) equipped with a DigiStar P2010 (NanoMEGAS, Brussels, Belgium) precession system. Data were recorded by a side entry Phurona CMOS (EMSIS, Münster, Germany) camera (12 M pixel model). An acceleration voltage of 200 kV was applied, and the camera length was set at 1200 mm. A 40 µm condenser aperture and a spot size of 3 were used, and the beam illuminating the sample had a diameter of 105 nm. A Gatan Elsa single-tilt cryo-transfer sample holder was used, enabling data acquisition at a temperature of 103 K. The sample was ground and deposited directly on a carbon-coated Cu grid without prior sonification of the powder in a solvent. In "precession tomography" mode, diffraction patterns were consecutively acquired by manually tilting (without the use of automated acquisition software) the same crystal along the α-axis of the single-tilt cryo-transfer sample holder while precessing the beam. The tilting step size was 1°, and the crystal was rotated in an angular range of −57° to +48°. During rotation, the electron beam was deviated to minimize the exposure time of the sample. Data reduction and integration were performed using PETS2.0 software [36].

We measured 14 crystals, which exhibited different degrees of diffuse scattering in the 00*l* direction of the reconstructed sections 0*kl* and *h*0*l* (data not shown). For the sake of simplicity, results from one crystal, which exhibited a negligible amount of diffuse scattering, will be discussed in the following. The crystal size used for the data collection had dimensions of ~400 × ~200 nm (see Figure S1), rendering it well-suited for the chosen method. During the data-collection process, the electron exposure duration (500 ms per step) was insufficient to cause observable degradation of the zeolite crystals. As a result, potential disruptions to the quality of the collected data were effectively minimized. The merging was carried out in the Laue class *mmm*, yielding R_{int} = 0.1837 for n_{obs} = 439 (n_{all} = 4428) reflections. The reconstructed reciprocal-space sections were calculated using PETS2, and sections *hk*0, 0*kl*, and *h*0*l* are shown in Figure 2a–c. Utilizing the Jana2020 software [37], the structural solution, derived through the charge-flipping algorithm applied to the extracted intensities, directly provided the atomic positions of silicon and oxygen atoms of the ECR-1 framework with **EON** topology. However, this process required additional steps, including successive difference Fourier map calculations and recycling, to precisely determine the positions of the extra-framework species (compensating Na^+ cations and water molecules). The refinement process included kinematic least-squares refinements on squared amplitudes, denoted as $|F|^2$, and was carried out using the Jana2020 software. To obtain the most accurate model from electron diffraction data, accounting for dynamical diffraction effects, a subsequent dynamical refinement was conducted. In accordance with the dynamical theory of diffraction, the experimentally observed intensities are influenced not only by structural factors but also by the crystal's orientation and thickness. This introduces complexity to the relationship between observed intensities and calculated structural factors. Various parameters must be incorporated during the refinement procedure to account for these effects. The Jana2020 software provides options for dynamical refinement, and the specific details of this method are reported elsewhere [22]. The illustrations of the structures were produced using Vesta [38] or Diamond [39] software.

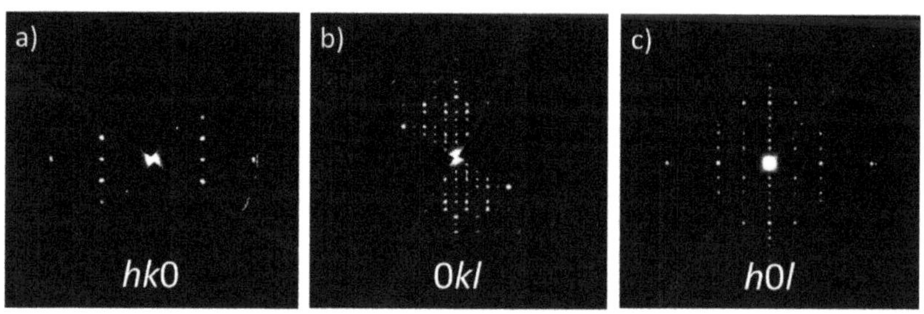

Figure 2. Reconstructed reciprocal-space sections along three different directions (**a**–**c**).

In order to investigate the distributions of sodium cations and water molecules throughout various crystallographic sites in the ECR-1 structure, we employed the Monte Carlo (MC) simulation technique using the procedure previously employed for potassium-exchanged faujasite zeolite (KX) [40]. The MC simulations were performed in the canonical ensemble (NVT), where the number of simulated particles N and the simulation cell volume V are kept fixed throughout the simulation, and temperature T is the temperature of the 3D ED experiment (103 K), applying the 8.0.0 version of the Towhee code [41]. The initial microscopic model for the ECR-1 zeolite was derived from the experimental structure with a unit cell chemical composition given by the formula $Na_{11}Si_{49}Al_{11}O_{120}$ and unit cell parameters of a = 7.4865 Å, b = 17.8463 Å, and c = 25.6545 Å. The aluminum atoms were then distributed according to the Loewenstein rule [42], and various starting Na^+ charge-compensating cation distributions were prepared.

The partial-charge distribution through the atoms of a semi-ionic zeolite structure are given in Table 1. To model the inter-molecular zeolite/zeolite, water/zeolite, and water/water interactions, we employed an expression containing a repulsion–dispersion term and an electrostatic potential term. These contributions were described using the Lennard–Jones (LJ) and Coulombic formulae, respectively. For the ECR-1 zeolite, we extracted the whole set of LJ and Coulombic parameters (q, ε, and σ) from the Clay force field [43]. For the water molecule, a five-point potential TIP5P model was employed [44]. In this model, which can reproduce the water liquid density, only oxygen atoms constitute LJ interaction sites. Moreover, it maintains the molecule rigid with the geometrical parameters set on the experimentally observed values in the gas phase (O-H bond length of 0.9572 Å and HOH angle of 104.52°). Finally, we used the Lorentz–Berthelot combination rules in order to determine the LJ interaction parameters for any pair of unlike atoms.

Table 1. The applied parameters of the Lennard–Jones interaction potential (ε, σ) and the partial atomic charges (q) distributed on atoms of adsorbent [43] and adsorbate [44] species.

Parameters for the Non-Bonded Interactions			
Atom	**ε/k_B (K)**	**σ (Å)**	**q (e)**
$H(H_2O)$	0	0	0.241
$O(H_2O)$	80.52	3.12	0
$L(H_2O)$ (lone pair interaction site)	0	0	−0.241
O (zeolite)	78.20	3.17	−1.094
Si (zeolite)	0.93×10^{-3}	3.30	2.10
Al (zeolite)	0.93×10^{-3}	3.30	1.58
Na^+	65.47	2.35	1.00

The simulation box consisted of the unit cell multiplied by four in the x-direction and two in the y-direction (29.9460 Å, 35.6926 Å, and 25.6545 Å) and contained 32 water molecules (corresponding to the experimental loading of 4 H_2O molecules per unit cell). During the MC simulations, the framework atoms (Si, Al, and O) were fixed, whereas the charge-compensating cations and the water molecules were able to displace. An equilibration phase of a Monte Carlo run consisted of 1.0×10^7 steps, while a production phase consisted of 2.0×10^7 steps. The following MC moves were employed: a cation translation and cation intra-box swap move, a water molecule center of mass translation, and a center of mass rotation and water molecule intra-box swap move [41]. The targeted acceptance rate of approximately 50% was maintained by regulating the maximum allowed translations and rotations, which were adjusted during the equilibration phase of the MC run. Periodic boundary conditions were applied in all three spatial directions. Ewald summation was employed to account for Coulombic interactions. Additionally, a cut-off distance of 12.5 Å was considered for Van der Waals interactions.

3. Results and Discussion

As demonstrated in Figure 3, the high-resolution SPDP acquired reveals the existence of trace amounts of nanosized mazzite and mordenite zeolites in the ECR-1 sample. Furthermore, consistent with the findings of Leonowicz and Vaughan [14], our sample displays stacking disorder, as evidenced by the results of the Le Bail refinement (Figure 3).

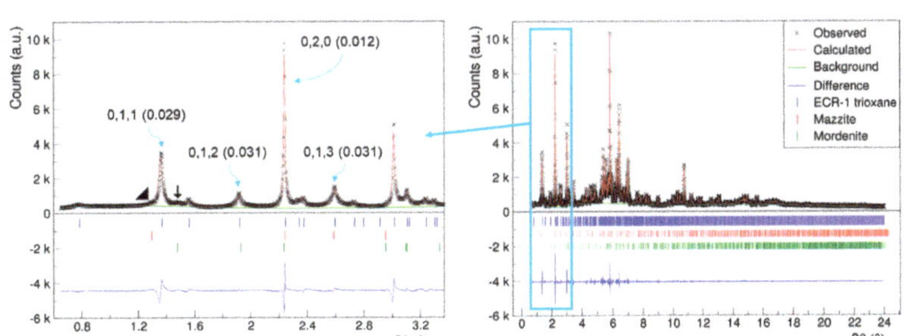

Figure 3. Le Bail refinement plot of ECR-1 (λ = 0.354196 (3) Å, a = 7.56159 (6) Å, b = 18.1016 (2) Å, c = 26.0313 (4) Å, wR = 4.8%) showing the presence of trace amounts of mazzite (MAZ)◢ and mordenite (MOR) ↓ zeolites. The Miller indices are indicated alongside the FWHM values in parentheses for representative reflections.

It is important to highlight that the observed peak broadening is not solely attributable to crystal size/shape effects; rather, there is an anisotropic peak broadening along the c-axis, similar to the findings in the structural study conducted by Gualtiri et al. [10]. The FWHM of the reflections showed this marked anisotropy, where reflections with $l \neq 0$ are broad and those with $l = 0$ are sharper. To accurately reproduce the observed powder pattern and account for stacking faults, a DIFFaX [45] simulation is imperative [10] in this case. Accordingly, we conducted a structural study using the 3D ED technique. Our objective was to isolate a nanocrystal with minimal stacking faults, allowing us to directly determine the structure of ECR-1 with **EON** topology. Furthermore, this approach to structural resolution from single-crystal data provides greater ease in locating compensation cations compared with powder-diffraction methods.

Following kinematical refinement, the final R_{obs} was determined to be 0.1690, and the refined composition was identified as $Na_{8.84}Si_{60}O_{120}$. Additional statistical indicators and refinement parameter details can be found in Table 2. In the kinematical structure refinement, all extra-framework species were refined as compensating cations without considering the presence of water molecules. This oversight was rectified in the dynamical refinement by examining the short Na…Na distances around 2.5 Å and replacing the corresponding sodium atoms with oxygen atoms of water molecules. The post-refinement composition was determined as $(Ow)_{4.52}Na_{8.16}Si_{60}O_{120}$ (where Ow represents the water molecule). Further details regarding the dynamical refinement parameters are shown in the right column of Table 2. The positional parameters, including x, y, and z coordinates, occupancies, and atomic displacement parameters, are provided in Tables S1 and S2 for the kinematical and dynamical refinements, respectively.

Table 2. Crystal and structural refinement data for ECR-1, with two distinct columns highlighting the values that differ between kinematical and dynamical refinements.

	Kinematical Refinement	Dynamical Refinement
Refined empirical formula	$Na_{8.84}Si_{60}O_{120}$	$Na_{8.16}Si_{60}O_{124.52}$
Formula weight	3802.0	3869.5
Temperature/K	103.0	
Crystal system	orthorhombic	
Space group	*Pmmn* (#59)	
a/Å	7.487 (3)	
b/Å	17.846 (12)	
c/Å	25.655 (8)	
$\alpha = \beta = \gamma/°$	90	
V/Å^3	3427.60 (15)	
Z	1	
ρ_{calc} g/cm^3	1.842	1.875
F(000)	629.896	635.553
Crystal size/nm	~400 × ~200	
Data collection	TEM Philips CM200,	
Collection mode	Precession-assisted 3D ED	
Source (wavelength)	electrons (λ = 0.02508 Å)	
2θ range for data collection/°	0.05 to 1.01	
Index ranges	$-8 \leq h \leq 9$, $-23 \leq k \leq 23$, $-31 \leq l \leq 26$	
Reflections collected	4367	17,481
Independent reflections	438 (R_{int} = 0.1837)	1040 (no averaging done)
Data coverage for $sin\theta/\lambda$ = 0.6 Å^{-1}	92.4%	93%
Data/restraints/constraints/parameters	4367/0/137	17481/17/32/275
Avg./Min/Max crystal thickness	-	526/25/2386 Å
g_{max}, R_{Sg}(max)	-	1.3, 0.5
Goodness-of-fit on F^2	obs: 2.34, all: 1.281	obs: 2.62, all: 0.7976
Final R values [$I \geq 3\sigma(I)$]	$R_{obs} = 0.169$, $wR_{2obs} = 0.358$	$R_{obs} = 0.097$, $wR2_{obs} = 0.1798$
Final R values [all data]	$R_{all} = 0.604$, $wR_{2all} = 0.641$	$R_{all} = 0.517$, $wR_{2all} = 0.2245$

For the treatment of 3D ED data, the persistence of the dynamical effects in kinematical refinement is reflected in the relatively high R values. This phenomenon is commonly observed in electron diffraction techniques, even extending to 3D ED methods, where R_{obs} values of approximately 0.2 or higher are routinely encountered during kinematical refinements [30,46–48]. However, for the dynamical refinements, the R_{obs} value substantially decreases, reaching as low as 0.097, as shown in Table 2. This fact confirms that high R values obtained during kinematical refinements are mainly due to dynamical effects. Another noteworthy point is the proximity of the atomic scattering factors for Si and Al. This proximity poses a challenge in refining Si and Al positions separately, particularly in the case of mixed tetrahedral site occupancies. Accordingly, in both kinematical and dynamical refinements, all tetrahedral atoms in the framework are refined as Si.

The dynamical refinement method yields superior figures of merit, provides a more comprehensive explanation of the observed electron diffraction intensities, and more accurately represents the crystal structure. As a result, the remainder of the discussion will focus on the outcomes derived from the dynamical refinement process.

As indicated in Table 2, the dynamical refinement results in a composition of $Na_{8.16}Si_{60}O_{124.52}$, where 4.52 oxygen atoms are from adsorbed water molecules. This chemical composition, in terms of sodium content, deviates from the average chemical composition of $Na_{10.40}Al_{10.40}Si_{49.60}O_{120}$, as determined by EDX (see Figure S2). Approximately 78.5% of the cations are located in the structure, distributed across six distinct crystallographic sites, with occupancies ranging from ~0.08 to ~0.8. The disparity between the chemical analysis of the powder and the result of the structural study on an individual nanocrystal can be attributed to the non-uniform distribution of aluminum atoms throughout the sample, potentially altering the number of negative charges within the framework that require compensation.

In a prior study, based on powder XRD data and using the Rietveld method, Gualtieri et al. identified different crystallographic sites for Na^+ in a hydrated zeolite ECR-1 with a similar chemical formula of $Na_{10.97}Ca_{0.36}(H_2O)_x[Al_{11.54}Si_{48.46}O_{120}]$ per unit cell [10]. Figure S3 illustrates the structure refined in this study and that of Gualtieri et al. [10]. The high vacuum conditions in the TEM sample chamber (~10^{-5} Pa) led to partial dehydration, as evidenced by the notable difference in the number of water molecules between the two structures: around 37.5 in the Rietveld study [10] compared with 4.5 in the present structural analysis using 3D ED. In the hydrated state of the ECR-1 sample, the observed weight loss on the thermogravimetric curve after dehydration is approximately 12%, corresponding to roughly 30 water molecules per unit cell [15]. The bond distances between both structures are reported in Table 1Sa in ref. [10] and in Table S3. The average Si—O bond lengths change significantly between the two framework structures, with averages of 1.64 Å in the Rietveld study and 1.61 Å in the current 3D ED study. The O-Si-O angles exhibit similar deformations of the silicon tetrahedrons, where they vary from 95.2° to 122.5° in the Rietveld study and from 101.4° to 122.1° in the present work after dynamical refinement.

Table 3 shows that both crystallographic studies reveal consistent interactions among the compensating Na^+ cations, water molecules, and the framework oxygen atoms. The average shortest distances recorded are 2.58 Å and 2.62 Å in the Rietveld study and the current 3D ED study, respectively. Notably, the Na…O distances exhibit a range, with minimum values of 2.25 Å and 2.29 Å and maximum values of 3.06 Å and 3.08 Å in the former Rietveld and current 3D ED crystallographic studies, respectively.

Table 3. Comparison of Na…O shortest distances after dynamical refinements at 103 K with the Rietveld analysis by Gualtieri et al. [10] at room temperature. Na^+ cations are labeled as C1 to C4.

	Gualtieri et al. [10]				This Work		
C1	O13	2×	2.25(4)	Na1	Ow3	2×	2.48(8)
	H_2O10	1×	2.36(9)		O22	1×	2.72(4)
	O8	2×	2.36(5)		O10	2×	2.98(2)
	O5	2×	2.95(7)	Na2	Ow2	1×	2.29(2)
	O11	1×	3.03(7)		Ow1	1×	2.32(3)
C2	O11	1×	2.28(4)		O2	2×	2.64(1)
	H_2O1	2×	2.65(2)		O13	1×	2.66(2)

Table 3. *Cont.*

	Gualtieri et al. [10]				This Work		
	O6	2×	2.76(2)		O9	1×	2.79(2)
	H_2O6	1×	3.06(5)		O16	2×	2.85(2)
C3	H_2O4	1×	2.35(2)	Na3	O14	2×	2.41(6)
	H_2O5	1×	2.49(4)		O11	2×	2.59(5)
	O20	1×	2.58(3)	Na4	O6	1×	2.46(6)
	O14	1×	2.62(3)		O14	2×	2.52(5)
C3b	O16	1×	2.40(4)		O1	2×	2.57(5)
	H_2O4	2×	2.47(4)	Na5	Ow2	1×	2.44(1)
	O19	2×	2.50(4)		O13	1×	2.73(1)
C4	H_2O7	1×	2.37(6)		O16	2×	2.83(8)
	H_2O3	2×	2.51(4)		O18	2×	3.08(1)
	O22	2×	2.954(4)	Na6	O14	2×	2.33(3)
					O7	1×	2.41(6)
					O11	2×	2.84(4)

However, there is variation in the distribution within the porosity, as depicted in Figure 4. In the hydrated ECR-1, as studied by Gualtieri et al. [10], 67.4% of the cations (C1, C3a, and C3b) are localized in the straight 12MRs channels (Figure 4a).

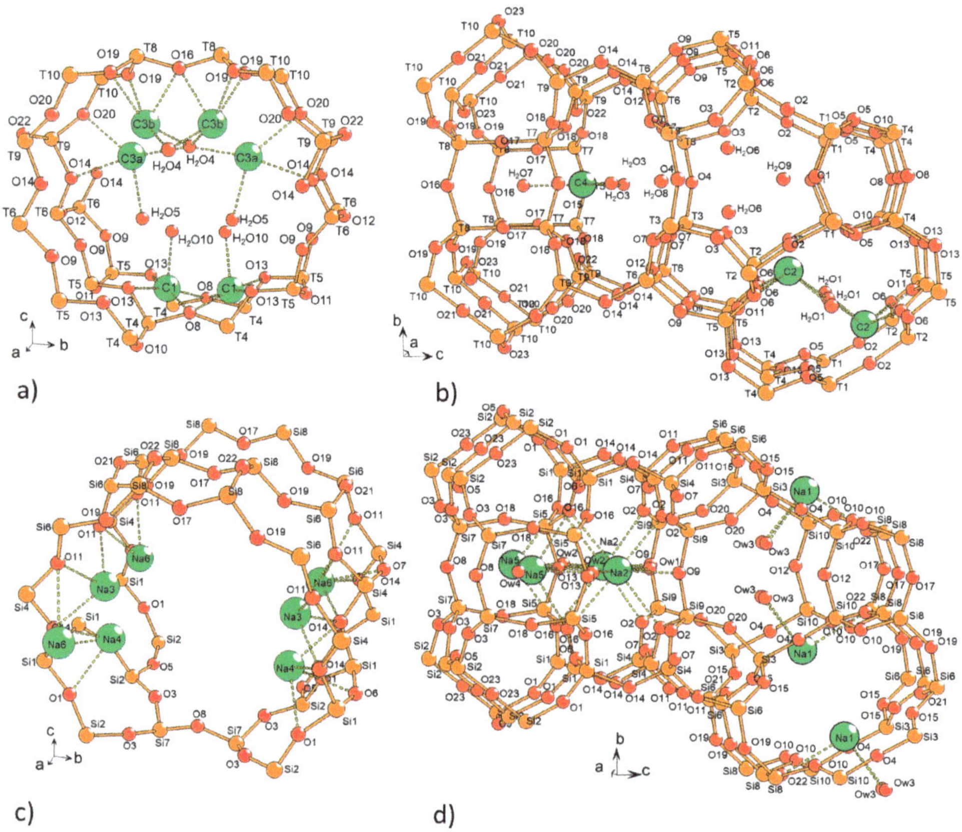

Figure 4. Perspectives showing the cation sites in ECR-1 (**a**,**b**) after the Rietveld analysis by Gualteri et al. [10], where Na^+ cations are labeled as C1 to C4, and (**c**,**d**) from the present crystallographic study using 3D ED. The interatomic distances (Na. . .O dashed bonds) between the cations and oxygen atoms of the framework, as well as the water molecules, are provided in Table 3.

For the remaining cations, 12.4% are distributed in the 8MRs channels (C4), connecting PerBU1 and PerBU2, and 20.2% within the 8MRs channels (C2) of PerBU2 (Figure 4b). In our present 3D ED study, after partial dehydration, only 32.4% of the cations are located inside the 12MRs (Na3, Na4, and Na6, Figure 4c). The other cations are situated as follows: 4.0% inside the 8MRs opening in PerBU1 (Na5); 38.7% in the 8MRs channels connecting PerBU1 and PerBU2 (Na2) in a site close to C4; and 24.9% within the 8MRs channels of PerBU2 (Na1) (Figure 4d) equivalent to C2. The water molecules remaining after partial dehydration under the vacuum of the TEM are predominantly confined in cavities with openings at 8MRs. This implies an accelerated and facilitated evacuation process when these molecules are situated within the 12MRs channels.

We initially anticipated the presence of trioxane molecules within the pores of ECR-1. Indeed, following the thermogravimetric analysis under air, Chatelard et al. proposed its location in the *gme* composite building units, possibly as a [Na^+-trioxane] complex [15]. However, Fourier difference maps did not reveal corresponding residual electron densities in the *gme* cage, in the straight 12MRs channels, or elsewhere. This outcome is attributed to the facile evacuation of the neutral trioxane molecules under vacuum. Nevertheless, it is not improbable that the electron beam has induced the reduction of trioxane molecules into carbon monoxide, which can be more readily evacuated through the 8MRs windows of the *gme* units than trioxane itself. Such damage from electron beams on zeolites is frequent, more specifically for the organic part [49].

The averaged cationic distributions of the experimentally determined crystallographic sites in both totally dehydrated and partially hydrated ECR-1 zeolite structures are summarized in Table 4. Comparing the simulated and experimental distributions reveals that cationic site occupancies exhibit similar behavior. Sites Na1 and Na2 are the most populated, followed by sites Na3, Na6, and Na4. In contrast, site Na5, which was unoccupied in the simulation, has the lowest experimental occupancy.

Table 4. Average computed number of charge-compensating Na^+ cations located proximal to the corresponding experimentally determined crystallographic sites per unit cell.

H_2O/Unit Cell		Na1	Na2	Na3	Na4	Na5	Na6	Others
0	Nb. of cations	3.05	2.79	1.50	0.03	0.00	0.49	3.14
	s.o.f. (MC@103 K) [a]	0.76	0.70	0.38	0.01	0.00	0.16	
4	Nb. of cations	2.41	3.00	1.13	0.16	0.00	0.35	3.95
	s.o.f. (MC@103 K) [a]	0.60	0.75	0.28	0.04	0.00	0.09	
	s.o.f. (3D ED) [b]	0.507	0.789	0.212	0.209	0.082	0.240	

[a] Site occupancy factors from the canonical Monte Carlo simulation; [b] site occupancies factor from the 3D ED crystallographic study.

Table 5 provides a comprehensive overview of the average distribution of water molecules across various crystallographic sites in ECR-1 zeolite. Despite the non-negligible probability of finding water molecules in the straight 12MRs channels, as depicted in Figure 5, the calculated locations around the cationic sites align well with the experimental trend. While a slight disparity exists between the simulated and experimental cationic distribution and the positioning of water molecules, it is crucial to acknowledge that the Monte Carlo simulation relied on a generic force field. Additionally, the distribution of aluminum atoms within the framework can impact cationic locations. Therefore, given the simplicity of the computational approach, the simulation demonstrates a commendable agreement with the experimental results.

Table 5. Average computed number of water molecules in experimentally determined crystallographic sites.

	Ow1	Ow2	Ow3	Ow4	Na5	Others
Nb. of H_2O molecules	0.50	0.38	0.89	0.13	0.32	2.10
s.o.f. (MC-103K) [a]	0.25	0.19	0.11	0.06	0.16	

[a] Site occupancy factors from the canonical Monte Carlo simulation.

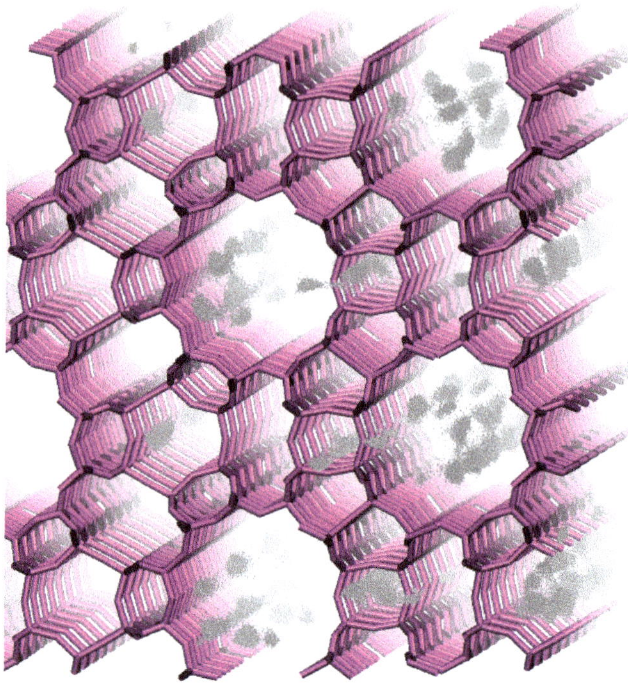

Figure 5. Perspective along the *a*-axis featuring 4000 superposed configurations of H_2O within the simulation box of ECR-1 zeolite at a loading of 4 H_2O per unit cell. Each grey dot corresponds to an oxygen atom of a water molecule.

4. Conclusions

The first structural analysis of ECR-1 zeolite using the 3D ED technique PED has revealed significant insights into its composition, crystal structure, and distribution of charge-compensating cations and water molecules within its framework. The final dynamical refinement (R_{obs} = 0.097) led to a chemical composition of $Na_{8.16}Si_{60}O_{124.52}$, with 4.52 oxygen atoms attributed to adsorbed water molecules. This composition differs from the average chemical composition determined by EDX, indicating a non-uniform distribution of aluminum atoms within the sample. Comparison with a prior Rietveld study highlights the impact of the dehydration phenomenon in the microscope on the crystal structure of ECR-1. The distribution of cations within the ECR-1 framework varies, with the remaining water molecules predominantly confined to cavities that connect 12MRs straight channels. The absence of OSDA trioxane used for the synthesis was attributed to its facile evacuation under the vacuum. However, the possibility of electron beam-induced damage cannot be disregarded. Overall, the experimental findings align well with simulated distributions of cations and water molecules, despite slight disparities that are attributed to the complexity of the framework and computational limitations. The study underscores the importance of advanced structural analysis techniques in elucidating the complex nature of zeolite frameworks and provides valuable insights for further understanding their properties and potential applications.

Supplementary Materials: The following supporting information can be downloaded at: https://www.mdpi.com/article/10.3390/sym16040477/s1, Figure S1: TEM image of the measured ECR-1 crystal. Figure S2: Energy dispersive X-ray spectroscopy (EDX) analyses of ECR-1 zeolite. Figure S3: Projections along the [100] plane of the ECR-1 zeolite (Gualtieri et al. [10] and this 3D ED study, dynamical refinement). Table S1: Fractional atomic coordinates and isotropic displacement parameters (Å^2) of ECR-1 (3D ED, kinematical refinement). Table S2: Fractional atomic coordinates and isotropic displacement parameters (Å^2) of ECR-1 (3D ED, dynamical refinement). Table S3: Si-O bond lengths and angles of ECR-1 (3D ED, dynamical refinement), a CIF file, and the corresponding checkCIF file.

Author Contributions: Conceptualization, J.-L.P.; investigation, T.Ö., I.D., C.C., M.D., A.T., R.M.-F. and J.-L.P.; writing—original draft preparation, T.Ö., I.D. and J.-L.P.; writing—review and editing, J.-L.P.; visualization, T.Ö., I.D. and J.-L.P.; supervision, J.-L.P. All authors have read and agreed to the published version of the manuscript.

Funding: This research received no external funding.

Data Availability Statement: CSD 2338838 contains the supplementary crystallographic data for this paper. These data can be obtained free of charge from FIZ Karlsruhe via www.ccdc.cam.ac.uk/structures. (accessed on 8 March 2024).

Acknowledgments: We express our heartfelt gratitude to Pierre Durand of Eloïse S.A.S. for his invaluable contributions in adapting the DigiStar P2010 (NanoMEGAS) precession system and his dedicated efforts in maintaining our CM200 microscope. Special thanks go to Loïc Benariac-Doumal, Catherine Dejoie, and Andrew Fitch from ESRF for their invaluable assistance in obtaining the high-resolution powder diffractogram on ID22. Additionally, we extend our appreciation to Laurent Legras, Gatan-EDAX Sales Manager in France, and Ludovic Josien from IS2M for the loan of the single-tilt cryo-transfer sample holder and the EDX measurement, respectively. The authors would like to acknowledge the High-Performance Computing Center of the University of Strasbourg for supporting this work by providing scientific support and access to computing resources.

Conflicts of Interest: The authors declare no conflicts of interest.

References

1. Li, Y.; Li, L.; Yu, J. Applications of zeolites in sustainable chemistry. *Chem* **2017**, *3*, 928–949. [CrossRef]
2. Li, Y.; Yu, J. Emerging applications of zeolites in catalysis, separation and host–guest assembly. *Nat. Rev. Mater.* **2021**, *6*, 1156–1174. [CrossRef]
3. Database of Zeolite Structures. Available online: http://www.iza-structure.org/databases/ (accessed on 15 December 2023).
4. Frising, T.; Leflaive, P. Extraframework cation distributions in X and Y faujasite zeolites: A review. *Microporous Mesoporous Mater.* **2008**, *114*, 27–63. [CrossRef]
5. Chao, C.C. Process for Separating Nitrogen from Mixtures Thereof with Less Polar Substances. U.S. Patent 4,859,217, 22 August 1989.
6. Vogt, E.T.C.; Weckhuysen, B.M. Fluid catalytic cracking: Recent developments on the grand old lady of zeolite catalysis. *Chem. Soc. Rev.* **2015**, *44*, 7342–7370. [CrossRef] [PubMed]
7. Vaughan, D.E.W.; Strohmaier, K.G. Crystalline Zeolite (ECR-1) and Process for Preparing It. U.S. Patent 4,657,748, 14 April 1987.
8. Database of Zeolite Structures—Building Scheme for EON. Available online: http://www.iza-structure.org/databases/ModelBuilding/EON.pdf (accessed on 15 December 2023).
9. Hsia Chen, C.S.; Schlenker, J.L.; Wentzek, S.E. Synthesis and characterization of synthetic zeolite ECR-1. *Zeolites* **1996**, *17*, 393–400. [CrossRef]
10. Gualtieri, A.F.; Ferrari, S.; Galli, E.; Di Renzo, F.; van Beek, W. Rietveld structure refinement of zeolite ECR-1. *Chem. Mater.* **2006**, *18*, 76–84. [CrossRef]
11. Song, J.; Dai, L.; Ji, Y.; Xiao, F.-S. Organic template free synthesis of aluminosilicate zeolite ECR-1. *Chem. Mater.* **2006**, *18*, 2775–2777. [CrossRef]
12. Lu, T.; Wang, Z.; Zhang, H.; Qin, J.; Yang, Y.; Cheng, P.; Zhao, Z. Radicalized seeds-assist route for the rapid synthesis of zeolite ECR-1 in the absence of organic templates. *Microporous Mesoporous Mater.* **2022**, *341*, 112071. [CrossRef]
13. Zhang, Y.J.; Chen, H.; He, P.Y.; Juan Li, C. Developing silica fume-based self-supported ECR-1 zeolite membrane for seawater desalination. *Mater. Lett.* **2019**, *236*, 538–541. [CrossRef]

14. Leonowicz, M.E.; Vaughan, D.E.W. Proposed synthetic zeolite ECR-1 structure gives a new zeolite framework topology. *Nature* **1987**, *329*, 819–821. [CrossRef]
15. Chatelard, C.; Dodin, M.; Martinez-Franco, R.; Tuel, A. Di- and trioxacyclohexane as structure directing molecules in the synthesis of zeolites omega and ECR-1. *Microporous Mesoporous Mater.* **2021**, *318*, 111015. [CrossRef]
16. Weirich, T.E.; Lábár, J.L.; Zou, X. *Electron Crystallography: Novel Approaches for Structure Determination of Nanosized Materials*; Springer: Dordrecht, The Netherlands, 2006; p. 536.
17. Vincent, R.; Midgley, P.A. Double conical beam-rocking system for measurement of integrated electron diffraction intensities. *Ultramicroscopy* **1994**, *53*, 271–282. [CrossRef]
18. Midgley, P.A.; Eggeman, A.S. Precession electron diffraction—A topical review. *IUCrJ* **2015**, *2*, 126–136. [CrossRef] [PubMed]
19. Mugnaioli, E.; Gorelik, T.; Kolb, U. "Ab initio" structure solution from electron diffraction data obtained by a combination of automated diffraction tomography and precession technique. *Ultramicroscopy* **2009**, *109*, 758–765. [CrossRef] [PubMed]
20. Wan, W.; Sun, J.; Su, J.; Hovmoller, S.; Zou, X. Three-dimensional rotation electron diffraction: Software RED for automated data collection and data processing. *J. Appl. Cryst.* **2013**, *46*, 1863–1873. [CrossRef] [PubMed]
21. Wang, B.; Zou, X.; Smeets, S. Automated serial rotation electron diffraction combined with cluster analysis: An efficient multi-crystal workflow for structure determination. *IUCrJ* **2019**, *6*, 854–867. [CrossRef] [PubMed]
22. Palatinus, L.; Petříček, V.; Corrêa, C.A. Structure refinement using precession electron diffraction tomography and dynamical diffraction: Theory and implementation. *Acta Crystallogr. Sect. A Found. Adv.* **2015**, *71*, 235–244. [CrossRef]
23. Gemmi, M.; Mugnaioli, E.; Gorelik, T.E.; Kolb, U.; Palatinus, L.; Boullay, P.; Hovmöller, S.; Abrahams, J.P. 3D electron diffraction: The nanocrystallography revolution. *ACS Cent. Sci.* **2019**, *5*, 1315–1329. [CrossRef] [PubMed]
24. Gruene, T.; Mugnaioli, E. 3D electron diffraction for chemical analysis: Instrumentation developments and innovative applications. *Chem. Rev.* **2021**, *121*, 11823–11834. [CrossRef]
25. Samperisi, L.; Zou, X.; Huang, Z. Three-dimensional electron diffraction: A powerful structural characterization technique for crystal engineering. *CrystEngComm* **2022**, *24*, 2719–2728. [CrossRef]
26. Klar, P.B.; Krysiak, Y.; Xu, H.; Steciuk, G.; Cho, J.; Zou, X.; Palatinus, L. Accurate structure models and absolute configuration determination using dynamical effects in continuous-rotation 3D electron diffraction data. *Nat. Chem.* **2023**, *15*, 848–855. [CrossRef]
27. Cho, J.; Willhammar, T.; Zou, X. The synergistic development of electron crystallography and zeolite discovery. *Microporous Mesoporous Mater.* **2023**, *358*, 112400. [CrossRef]
28. Lorgouilloux, Y.; Dodin, M.; Mugnaioli, E.; Marichal, C.; Bats, N.; Caullet, P.; Kolb, U.; Paillaud, J.-L. IM-17: A new zeolitic material, synthesis and structure elucidation from electron diffraction ADT data and Rietveld analysis. *RSC Adv.* **2014**, *4*, 19440–19449. [CrossRef]
29. Cichocka, M.O.; Lorgouilloux, Y.; Smeets, S.; Su, J.; Wan, W.; Caullet, P.; Bats, N.; McCusker, L.B.; Paillaud, J.-L.; Zou, X. Multidimensional disorder in zeolite IM-18 revealed by combining transmission electron microscopy and X-ray powder diffraction analyses. *Cryst. Growth Des.* **2018**, *18*, 2441–2451. [CrossRef]
30. Steciuk, G.; Schäf, O.; Tortet, L.; Pizzala, H.; Hornfeck, W.; Palatinus, L.; Paillaud, J.-L. A new lithium-rich zeolitic 10-MR zincolithosilicate MZS-1 hydrothermally synthesized under high pressure and characterized by 3D electron diffraction. *Eur. J. Inorg. Chem.* **2021**, *2021*, 628–638. [CrossRef]
31. Krysiak, Y.; Barton, B.; Marler, B.; Neder, R.B.; Kolb, U. Ab initio structure determination and quantitative disorder analysis on nanoparticles by electron diffraction tomography. *Acta Crystallogr. Sect. A Found. Adv.* **2018**, *74*, 93–101. [CrossRef]
32. Krysiak, Y.; Marler, B.; Barton, B.; Plana-Ruiz, S.; Gies, H.; Neder, R.B.; Kolb, U. New zeolite-like RUB-5 and its related hydrous layer silicate RUB-6 structurally characterized by electron microscopy. *IUCrJ* **2020**, *7*, 522–534. [CrossRef]
33. Fitch, A.; Dejoie, C.; Covacci, E.; Confalonieri, G.; Grendal, O.; Claustre, L.; Guillou, P.; Kieffer, J.; de Nolf, W.; Petitdemange, S.; et al. ID22—The high-resolution powder-diffraction beamline at ESRF. *J. Synchrotron Radiat.* **2023**, *30*, 1003–1012. [CrossRef]
34. Le Bail, A.; Duroy, H.; Fourquet, J.L. Ab-initio structure determination of $LiSbWO_6$ by X-ray powder diffraction. *Mater. Res. Bull.* **1988**, *23*, 447–452. [CrossRef]
35. Toby, B.H.; Von Dreele, R.B. GSAS-II: The genesis of a modern open-source all purpose crystallography software package. *J. Appl. Cryst.* **2013**, *46*, 544–549. [CrossRef]
36. Palatinus, L.; Brázda, P.; Jelínek, M.; Hrdá, J.; Steciuk, G.; Klementová, M. Specifics of the data processing of precession electron diffraction tomography data and their implementation in the program PETS2.0. *Acta Crystallogr. Sect. B Struct. Sci. Cryst. Eng. Mater.* **2019**, *75*, 512–522. [CrossRef] [PubMed]
37. Petříček, V.; Palatinus, L.; Plášil, J.; Dušek, M. Jana2020—A new version of the crystallographic computing system Jana. *Z. Kristallogr. Cryst. Mater.* **2023**, *238*, 271–282. [CrossRef]
38. Momma, K.; Izumi, F. VESTA 3 for three-dimensional visualization of crystal, volumetric and morphology data. *J. Appl. Cryst.* **2011**, *44*, 1272–1276. [CrossRef]
39. Brandenburg, K.; Putz, H. *Diamond—Crystal and Molecular Structure Visualization*; Version 3.2k4; Crystal Impact: Bonn, Germany, 2018; Available online: https://www.crystalimpact.de/diamond (accessed on 3 December 2023).
40. Ammouli, T.; Paillaud, J.-L.; Nouali, H.; Stephan, R.; Hanf, M.-C.; Sonnet, P.; Deroche, I. Insights into water adsorption in potassium-exchanged X-type faujasite zeolite: Molecular simulation and experiment. *J. Phys. Chem. C* **2021**, *125*, 19405–19416. [CrossRef]

41. Martin, M.G. MCCCS Towhee: A tool for Monte Carlo molecular simulation. *Mol. Simul.* **2013**, *39*, 1212–1222. [CrossRef]
42. Loewenstein, W. The distribution of aluminum in the tetrahedra of silicates and aluminates. *Am. Mineral.* **1954**, *39*, 92–96. Available online: https://pubs.geoscienceworld.org/msa/ammin/article-abstract/39/1-2/92/539331 (accessed on 21 October 2023).
43. Cygan, R.T.; Liang, J.-J.; Kalinichev, A.G. Molecular models of hydroxide, oxyhydroxide, and clay phases and the development of a general force field. *J. Phys. Chem. B* **2004**, *108*, 1255–1266. [CrossRef]
44. Mahoney, M.W.; Jorgensen, W.L. A five-site model for liquid water and the reproduction of the density anomaly by rigid, nonpolarizable potential functions. *J. Chem. Phys.* **2000**, *112*, 8910–8922. [CrossRef]
45. Treacy, M.M.J.; Newsam, J.M.; Deem, M.W.; Treacy, M.M.J.; Newsam, J.M.; Deem, M.W. A General recursion method for calculating diffracted intensities from crystals containing planar faults. *Proc. Math. Phys. Sci.* **1991**, *433*, 499–520. [CrossRef]
46. Kolb, U.; Gorelik, T.E.; Mugnaioli, E.; Stewart, A. Structural characterization of organics using manual and automated electron diffraction. *Polym. Rev.* **2010**, *50*, 385–409. [CrossRef]
47. Palatinus, L.; Jacob, D.; Cuvillier, P.; Klementova, M.; Sinkler, W.; Marks, L.D. Structure refinement from precession electron diffraction data. *Acta Crystallogr. Sect. A Found. Cryst.* **2013**, *69*, 171–188. [CrossRef] [PubMed]
48. Ge, M.; Wang, Y.; Carraro, F.; Liang, W.; Roostaeinia, M.; Siahrostami, S.; Proserpio, D.M.; Doonan, C.; Falcaro, P.; Zheng, H.; et al. High-throughput electron diffraction reveals a hidden novel metal–organic framework for electrocatalysis. *Angew. Chem. Int. Ed.* **2021**, *60*, 11391–11397. [CrossRef] [PubMed]
49. Girelli Consolaro, V.; Rouchon, V.; Ersen, O. Electron beam damages in zeolites: A review. *Microporous Mesoporous Mater.* **2024**, *364*, 112835. [CrossRef]

MDPI AG
Grosspeteranlage 5
4052 Basel
Switzerland
Tel.: +41 61 683 77 34

Symmetry Editorial Office
E-mail: symmetry@mdpi.com
www.mdpi.com/journal/symmetry

www.ingramcontent.com/pod-product-compliance
Lightning Source LLC
Chambersburg PA
CBHW062100220926
51850CB00010B/2831
* 9 7 8 3 7 2 5 8 5 3 6 1 8 *